Cellular Radio – Principles and Design

Cellular Radio

Principles and Design

R. C. V. Macario

Department of Electrical and Electronic Engineering
University of Wales, Swansea

McGraw-Hill, Inc.

New York   San Francisco   Washington, D.C.   Bogotá
Caracas   Mexico City   San Juan   Toronto

3 4 5 6 7 8 9 0   DOC/DOC   9 9 8 7 6 5

ISBN 0-07-044301-7

First published 1993 by THE MACMILLAN PRESS LTD
Houndmills, Basingstoke, Hampshire RG21 2XS
and London.

Printed and bound by R. R. Donnelley & Sons Company.

 This book is printed on recycled, acid-free paper containing a minimum of 50% recycled de-inked fiber.

Contents

Preface   ix

Abbreviations and Acronyms   xi

1 Introduction   **1**
   1.1 The radiotelephone   1
   1.2 Expanding the number of subscribers   6
   1.3 The cellular principle   7
   1.4 Radio coverage by a single cell   8
   1.5 Multiple cell layout   11
   1.6 The fixed supporting network   13
   1.7 Radio frequencies available   16
   1.8 The radio carrier and some attributes   18
   1.9 Control and channel signalling   22
   1.10 Error correction strategies   26
   1.11 Numbering plans   30
   1.12 Summary of important features   34

2 Radio Coverage Prediction   **37**
   2.1 Electromagnetic waves   37
   2.2 Antenna considerations   38
   2.3 Models for propagation   41
       2.3.1 *In free space*   42
   2.4 Reflections at a boundary   45
   2.5 Terrestrial propagation   47
       2.5.1 *Simple flat earth model*   47
       2.5.2 *Rough ground model*   50
       2.5.3 *CCIR standard model*   51
   2.6 Cell site coverage assessment   53
   2.7 Computer prediction techniques   53

3   Cellular Radio Design Principles   59
3.1 Analog cellular frequency allocation plans   59
3.2 Base station site engineering   62
3.3 The concept and benefits of channel sharing   62
3.4 Multiple cell plan   66
    3.4.1 *Cell structure geometry*   67
    3.4.2 *Reuse distance*   68
    3.4.3 *Cell splitting*   72
    3.4.4 *Sectorization*   73
    3.4.5 *Other cell patterns*   73
3.5 The cellular system   75
    3.5.1 *Mobile location*   75
    3.5.2 *In call handover*   76
3.6 The cellular network   78
    3.6.1 *Base stations*   79
    3.6.2 *Mobile switching centres*   80
    3.6.3 *Other services*   80

4   Analog Cellular Radio Signalling   82
4.1 Channel trunking needs   82
4.2 Equipment identity numbers   85
4.3 Radio link signalling details   86
    4.3.1 *Forward control channel messages*   87
    4.3.2 *Overhead messages*   87
    4.3.3 *Mobile station control messages*   89
    4.3.4 *Control filler messages*   90
4.4 Registration   90
    4.4.1 *Reverse control channel messages*   90
4.5 Mobile call initiation   92
    4.5.1 *Reverse control channel formatting*   92
4.6 Mobile call reception   93
4.7 The signalling tone   95
4.8 The supervisory audio tone   95
4.9 Handover   96
    4.9.1 *Illustration of signalling procedures*   97
4.10 Data over cellular   100
    4.10.1 *Data specific networks*   100

5	**The Multipath Propagation Problem**	**104**
	5.1 General considerations	104
	5.2 Multipath fading characteristics	104
	5.2.1 *Elementary multipath*	107
	5.2.2 *A scattering model*	108
	5.2.3 *Effect of vehicle velocity*	111
	5.2.4 *Fading envelope statistics*	112
	5.3 Diversity reception	114
	5.4 Frequency selective fading	115
	5.5 Coherence bandwidth and delay spread	116
6	**Modulation Techniques**	**119**
	6.1 Introduction	119
	6.2 The bandwidth problem	119
	6.3 Analog modulation bandwidths	121
	6.4 Shift key modulations	126
	6.4.1 *Phase shift keying*	126
	6.4.2 *Frequency shift keying*	128
	6.4.3 *Quadrature phase shift keying*	133
	6.4.4 *Minimum shift keying*	134
	6.4.5 *Tamed frequency shift keying*	135
	6.4.6 *Gaussian minimum shift keying*	135
	6.4.7 *Differential phase shift keying*	138
	6.5 Bit error rate	140
	6.5.1 *Improving BER*	144
7	**Speech Coding**	**147**
	7.1 Introduction	147
	7.2 Coding requirements	147
	7.3 Coding techniques	149
	7.3.1 *Waveform coders*	149
	7.3.2 *Vocoders*	150
	7.3.3 *Hybrid coders*	152
	7.3.4 *Codebook vocoders*	156

8   Digital Cellular Designs   **159**
     8.1 Second generation networks   159
     8.2 Time division multiple access   160
          8.2.1 *Possible advantages of TDMA*   163
     8.3 The pan-European cellular system   165
          8.3.1 *Features of GSM*   166
          8.3.2 *The OSI reference model*   167
          8.3.3 *The fixed network supporting GSM*   172
          8.3.4 *The radio part*   174
          8.3.5 *The timing structure of GSM*   176
          8.3.6 *Channel coding and training sequence*   179
          8.3.7 *Radio link management*   180
          8.3.8 *Signalling within GSM*   183
     8.4 The North American digital cellular system   186
          8.4.1 *Radio transmission strategy*   186
          8.4.2 *Control channels*   188

9   Spectral Efficiency Considerations   **191**
     9.1 Introduction   191
     9.2 Network example   192
     9.3 Bandwidth limit to subscribers   194
     9.4 Measures of spectral efficiency   195
          9.4.1 *Definition of traffic intensity*   196
          9.4.2 *Erlangs and unit calls*   197
     9.5 Grade of service   198
          9.5.1 *Telephone traffic formulas*   198
          9.5.2 *Activity in a cell*   200
     9.6 Calculation of spectral efficiency   201
          9.6.1 *Conventional FDMA*
     9.7 Multi-access efficiency factor
          9.7.1 *Overall efficiency*

Appendix: Summary of Current World Systems   209

Index
                                                                   211

Preface

This book is designed to appeal to any student of the technology and operation of cellular radio, whether at advanced undergraduate or postgraduate level, or undertaking a further training course, or at practitioner level. It is assumed that the reader has a basic knowledge of electronic engineering, and some experience in quantitative evaluation. The aim is to provide such a reader with an understanding of some of the most advanced person-to-person communication systems being installed in the world today.

The material included is drawn from experience gained in teaching courses on cellular communications to undergraduates and postgraduates at the University of Swansea, over several years, as well as from practical experience in designing, evaluating and operating person-to-person communication equipment.

An important first objective has been to set out the basic principles that underlie all cellular radio; these are explained in the introductory chapter. There are also design rules and strategies, and these are demonstrated by description of established and newly installed systems. While the physical realities of all aspects of cellular radio may be found in the text, their interpretation is left to the reader, whether a potential user, or designer, or system operator.

Having spent much of his career to date in the field of radio communication, the writer readily acknowledges the contribution made over the years to his teaching and research by data and information disseminated through seminars, especially those held by the leading manufacturers of test equipment, for the design and evaluation of radio equipment and cellular. In particular he wishes to acknowledge Hewlett Packard Company, Marconi Instruments Ltd, and Rohde & Schwarz UK Ltd. He would also like to record his indebtedness to the Engineering Division of Vodafone Ltd, directed by Ted Beddoes, managed by Roger Wilkins, and in particular to Melanie Foster who prepared the text manuscript at the Vodafone offices in Newbury. He would also like to thank Angela McGairl in Swansea for preparing the drawings. The above facts perhaps demonstrate that the writer had to be mobile to get the work accomplished, which perhaps has enhanced the mobility aspect of the subject.

The author is grateful to Dr Paul Lynn for his encouragement and guidance during the long period of preparation of the text, and also to Malcolm Stewart of the Macmillan Press Ltd for his exacting demands with regard to layout and presentation.

The writer is also grateful to the many students who have enthusiastically attended his lectures on this subject over the years. Many were from overseas centres, and as is always true in teaching, the teacher can learn much from his students. I am also grateful to my family for accepting that the manuscript took precedence at home and weekends for a long time, and in particular to my wife, Margaret for her continual support and encouragement.

R.C.V. Macario

Abbreviations and Acronyms Associated with Cellular Radio

AB	*Access Burst*
ACCH	*Associated Control Channel*
ACS	*Adjacent Channel/Carrier Suppression*
ACU	*Antenna Combination Unit*
ADC	*American Digital Cellular*
AFC	*Automatic Frequency Control*
AGC	*Automatic Gain Control*
AGCH	*Access Grant Channel*
AGWN	*Additive White Gaussian Noise*
AI	*Area Identification (field)*
AMPS	*Automatic Mobile Phone Service (USA)*
APC	*Airborne Public Correspondence (USA)*
ARQ	*Automatic Request for Retransmission*
AuC	*Authentication Centre*
Au	*Authentication*
BCC	*Base (station) Colour Code*
BCCH	*Broadcast Control Channel*
BCH	*Broadcast Channel*
BCH	*A particular error correcting code*
BER	*Bit Error Rate*
Bm	*Traffic channel for full-rate voice coder (ISDN terminology for mobile service)*
BS	*Base Station*
BSC	*Base Station Controller*
BSCU	*Base Station Controller Unit*
BSI	*Base Station Interface*
BSIC	*Base Station Identity Code*
BSS	*Base Station System*
BSSAP	*Base Station Application Part*
BTS	*Base Transceiver Station*
CA	*Cell Allocation*
CA-CN	*Cell Allocation RF Channel Number*
CBCH	*Cell Broadcast Channel*
CC	*Country Code*
CCCH	*Common Control Channel*
CHAN	*Voice Channel*

CI	*Cell Identity*
CPA	*Combined Paging Access (field)*
CRC	*Cyclic Redundancy Check*
CSPDN	*Circuit Switched Public Data Network*
CU	*Central Unit (of a MS)*
C/I	*Carrier-to-Interference Ratio*
D	*Downlink*
DB	*Dummy Burst*
DCC	*Digital Colour Code*
DCF	*Data Communication Function*
DCCH	*Dedicated Control Channel*
DCN	*Data Communication Network*
DL	*Data Link (layer)*
DLD	*Data Link Discriminator*
Dm	*Control Channel*
	(ISDN terminology for mobile service)
DMR	*Digital Mobile Radio*
DP	*Dialled Pulse*
DTAP	*Direct Transfer Application Part*
DTE	*Data Terminal Equipment*
DTMF	*Dual-Tone Multi-Frequency (signalling)*
DRX	*Discontinuous Reception*
DTX	*Discontinuous Transmission (field)*
Eb/No	*Ratio of energy per bit to noise power spectral density*
EIR	*Equipment Identity Register*
END	*End Indication (field)*
ESN	*Electronic Serial Number*
E-TACS	*Extended TACS (more channels)*
FA	*Full Association*
FB	*Frequency (correction) Burst*
FACCH	*Fast Associated Control CHannel*
FACCH/F	*Full-rate FACCH*
FACCH/H	*Half-rate FACCH*
FDMA	*Frequency Division Multiple Access*
FEC	*Forward Error Correction*
FER	*Frame Erasure Rate*
FCC	*Forward Control Channel*
FN	*Frame Number*

FREG	*Forced Registration (field)*
FVC	*Forward Voice Channel*
G	*Guard (bits)*
GIM	*Group Identification Mark*
GMSK	*Gaussian Minimum Shift Keying*
GSM	*Global System for Mobile Communications*
	previously: Group Special Mobile
GSM PLMN	*GSM Public Land Mobile Network*
HDLC	*High Level Data Link Control*
HLR	*Home Location Register*
HMSC	*Home Mobile Switching Centre*
HPLMN	*Home PLMN*
HPU	*Handportable Unit*
ID	*Identification*
IDN	*Integrated Digital Network*
IMSI	*International Mobile Subscriber Identity*
ISDN	*Integrated Services Digital Network*
IWF	*Inter Working Function*
JDC	*Japanese Digital Cellular*
J-TACS	*Japanese TACS system*
Ke	*Cipher Key*
LAC	*Location Area Code*
LAI	*Location Area Identity*
LAN	*Local Area Network*
LE	*Local Exchange*
Lm	*Traffic channel with capacity lower than Bm*
LPC	*Linear Predictive Coding (Voice Coder)*
LR	*Location Register*
MA	*Mobile Allocation*
MACN	*Mobile Allocation Channel Number*
MAP	*Mobile Application Part*
MCC	*Mobile Country Code*
MIC	*Mobile Interface Controller*
MIN	*Mobile Identification Number*

MMI	*Man Machine Interface*
MNC	*Mobile Network Code*
MS	*Mobile Station*
MSC	*Mobile-services Switching Centre*
MSCU	*Mobile Station Control Unit*
MS ISDN	*Mobile Station ISDN Number*
MSL	*Main Signalling Link*
MSRN	*Mobile Station Roaming Number*
NAM	*Number Assignment Module*
N-AMPS	*Narrowband AMPS system*
NB	*Normal Burst*
NCELL	*Neighbouring (adjacent) Cell*
NDC	*Network Destination Code*
NE	*Network Element*
NMC	*Network Management Centre*
NMSH	*National Mobile Station Identification (number)*
NMT	*Nordic Mobile Telephone (system)*
NSAP	*Network Service Access Point*
N(S)N	*National (significant) Number*
NT	*Network Termination*
OHD	*Overhead Message Type Field*
OMC	*Operations & Maintenence Centre*
OSI	*Open System Interconnection*
P	*Parity Field*
PCH	*Paging Channel*
PCM	*Pulse Code Modulator*
PDN	*Public Data Networks*
PIN	*Personal Identification Number*
PLMN	*Public Land Mobile Network*
PSPDN	*Public Switched Public Data Network*
PSTN	*Public Switched Telephone Network*
PTO	*Public Telecommunications Operator*
RCC	*Reverse Control Channel*
REC	*RECommendation*
REGH	*Registration Field for Mobile*
REL	*RELease*
RELP	*Residual Excited Linear Predictive (coder)*

REQ	*REQuest*
RES	*RESponse (authentication)*
RFCH	*Radio Frequency CHannel*
RLP	*Radio Link Protocol*
RPE	*Regular Pulse Excitation (Voice Coder)*
RSE	*Radio System Entity*
RSVD	*Reserved for Future Use (bits)*
RVC	*Reverse Voice Channel*
RX	*Receiver, or RX*
SAP	*Service Access Point*
SAPI	*Service Access Point Indicator*
SAT	*Supervisory Audio Tone*
SB	*Synchronization Burst*
SCC	*SAT colour code*
SCCP	*Signalling Connection Control Part*
SCH	*Synchronization CHannel*
SCM	*Station Class Mark*
SDCCH	*Stand-alone Dedicated Control CHannel*
SIDH	*System Identification of Home Mobile Service Area*
SIM	*Subscriber Identity Module*
SLTM	*Signalling Link Test Message*
SMS	*Short Message Service*
SN	*Subscriber Number*
SP	*Signalling Point*
ST	*Signalling Tone*
S/No	*Serial Number*
TA	*Terminal Adaptor*
TACS	*Total Access Communications (UK)*
TB	*Tail Bits*
TC	*Trunk Code*
TCH	*Traffic CHannel*
TCH/F	*Full-rate TCH*
TCH/H	*Half-rate TCH*
TCH/FS	*Full-rate Speech TCH*
TCH/HS	*Half-rate Speech TCH*
TCH	*Transceiver Control Interface*
TDD	*Time Division Duplex*
TDMA	*Time Division Multiple Access*
TE	*Terminal Equipment*

TFTS	*Terrestrial Flight Telephone System*
TMN	*Telecommunications Management Network*
TMSI	*Temporary Mobile Subscriber Identity*
TN	*Time Slot Number*
TRX	*Transceiver*
TS	*Time Slot*
TSC	*Training Sequence Code*
TX	*Transmitter, or TX*
U	*Uplink*
VAD	*Voice Activitity Detection*
VLR	*Visitor Location Register*
VMAC	*(Voice) Mobile Attenuation Code*

[Note: this list is not necessarily exhaustive]

Cellular Radio – Principles and Design

1 Introduction

Cellular radio is a complex technological system. It embraces several disciplines of engineering and has taken much enterprise and development to assemble into global systems. For example, cellular radio requires the combining of many large scale technologies, such as efficient high frequency semiconductor technology, radio transmission planning and global fixed telecommunications networks. The approach taken in this text, therefore, is firstly to set out an overview of the main disciplines and designs involved. Chapter 1 therefore presents this overview; the varied details of the many 'parts' can then be studied in the chapters which follow.

A particular aspect of cellular radio, as indeed is the case with many other subjects, is the appearance and common use of acronyms. Those generally found in use for cellular radio are listed at the front of the book. They are all abbreviations, for example, MSC - mobile switching centre, and there is a temptation to proceed using these abbreviations freely from then on in the text. This has been avoided where possible, at least in the earlier chapters, so that they can be read more easily. Towards the end of the text, however, so much detail is involved that these acronyms are used more often. By then the first time reader will have come to terms with many of the abbreviations.

Another attribute of cellular radio is that it is a mature technology, even though this maturity has come about in a short time scale; in fact, it is not much more than one decade old. The speed of realisation is much faster than that of other public domain engineering activities, such as the rail network, the fixed telephone network and motorway networks.

Although it is generally recognised that fixed telecommunications networks (telephones, facsimile, etc), are the largest and most completely integrated technological systems at present found in the world, cellular radio is fast becoming of equal size and complexity. In analogous terms, one could say that cellular radio is like the sunflower in nature; it begins with quite a small seed, grows with a tree-like stem within a season and produces a brilliant coloured flower - which then turns to face the sun (in cellular, the subscriber) - the sunflower also produces more of its own kind (seeds) in a cellular arrangement; cellular radio virtually does the same.

1.1 The radiotelephone

A radiotelephone can be defined as a telephone without wires, the connection to the local exchange being through the medium of *radio*. To achieve this operation many new factors above and beyond the basic telephone must be introduced.

Let us just recall what are the features of the so-called plain old telephone. Figure 1.1 shows these in a semi-descriptive way.

The telephone has a number which is registered solely in the local exchange (LE). Numbers can be dialled from the phone using a keypad, and the accepted dual-tone multi-frequency (DTMF) format, Figure 1.2, is the usual means of transferring these numbers. Power to operate the ringer and other functions in the phone is supplied by the local exchange through the two wires of the local loop. Only two wires are used, but these achieve a two-way or duplex mode of conversation. However, the length of the local loop to the exchange is limited, because amplification is not possible both ways with only two wires.

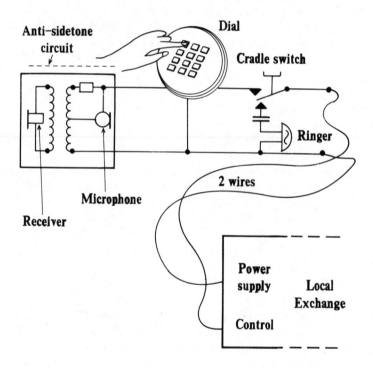

Figure 1.1   The basic telephone with a two-wire connection to the local exchange

Another feature is the cradle found on all telephones; this alerts the local exchange as to whether the phone is *on* or *off hook*. Such artifacts, which make the

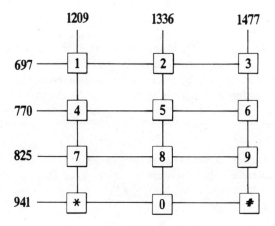

Figure 1.2   The dual-tone multi-frequency keypad and associated audio tones (two out of one-from-four code); shown in Hertz

telephone system work, are merely listed at this point in the description as features. These features are necessary within the fixed telephone network system, however.

Where does the *radiotelephone* (R/T) differ? Again a semi-descriptive diagram is a good place to start and here Figure 1.3 is perhaps helpful. Clearly a radiotelephone differs in many ways.

- An R/T requires a portable source of power - its (rechargeable) battery - in order to function.

- The local exchange has been replaced by a (local) *base station* (BS). The base station is fixedwire connected to a mobile telephone network - which may be part of, or in addition to, the fixed network mentioned above.

- Both the radiotelephone - now called the *mobile station* (MS) (but we will use the word mobile in general) - and the base station need a radio antenna.

- These antennas must be suitable for the radio frequencies which are allocated within the radio spectrum, being those licensed for use by the radio telephones in operation.

- Two radio channels in general must be allocated to each mobile phone in order to have duplex operation, that is, the user can speak and listen at the same time. Variation on this theme using time division multiplex operation is described later, but whatever the case, a *forward* and a *return* path radio channel is required.

The forward channel refers to the base-to-mobile path; the reverse channel refers to the mobile-to-base path. Both the BS and the MS require radio transmitter circuits. The weakest path is the return path, because the mobile unit has limited radio power, in order to conserve battery power and extend operational time between charging (overnight). Cellular radio is designed to overcome this limitation.

• The mobile carries its own telephone number in internal memories. This is unique to a personal radiotelephone or cellular radio. It allows the subscriber to roam over the network - provided that he can set up agreement with the local base station where he is operating.

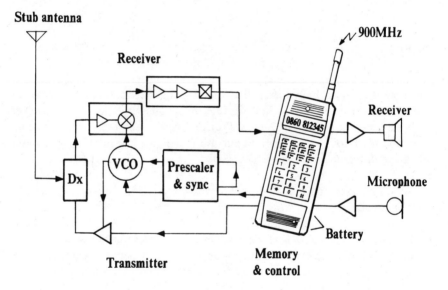

Figure 1.3   Outline of the components within a cellular radiotelephone

• The mobile will also contain a radio receiver, transmitter and tuning (synthesizer) circuits which must be supplied by power from the battery and which take and give instructions to the local memory and control module.

• The ringer is now controlled through the previous circuits. Notice that no cradle is found on a radiotelephone (as opposed to the much simpler cordless telephones which we do not discuss). The BS and MS in fact automatically keep in touch by various handshaking protocols. A call is set up by depressing a specific SEND button on the keyboard, not associated with the DTMF keyboard instructions.

It is clear that a radio or cellular telephone handset is completely different in technology from a fixed telephone instrument, not just in the fact that the wire to the local exchange is missing. The sets also differ specifically as regards several operational features.

- Provided that a mobile radio telephone network and service has been set up, by the network operator and the holder of the radio licence to run such a service, then the subscriber (to the service, the user of the radiotelephone) can *roam* around the network, or country as he wishes, so long as he stays within radio distance of a base station.

- Our subscriber has a telephone number, now called a mobile telephone number, which is registered to the handset. The number will usually be of the same format as numbers allocated to fixed subscribers, shown in general in Figure 1.4, except that the area or office code will not refer to a specific town or district say, but to a specific mobile telephone service.

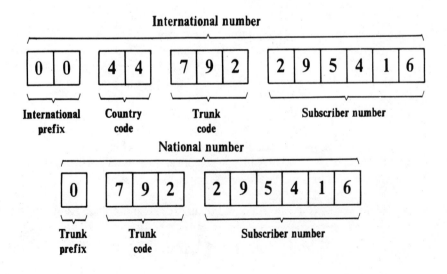

Figure 1.4   The international and national number format at the present time

Some examples of mobile service access codes in different parts of the world are shown here. (There is often more than one network operator.)

Australia	(0)18, (07), . . .
Finland	(0)49, . . .
Hong Kong	(0)903, (0)948, . . .
Sweden	(0)10, . . .
United Kingdom	(0)836, (0)860, . . .
United States	(0)516, (0)813, . . .

The fixed network will hold a register of mobile numbers for billing, authentication and location purposes, but the mobile phone number personal to the handset does of course go hand-in-hand with the roaming feature.

• Finally, for call management, much more sophisticated signalling between the handset and the local base station (and supporting network) is necessary, and specific instructions must be given by the user (by his keyboard usually). Thus a large amount of digital data traffic is found in cellular radio, whether the system is analog or digital.

1.2   Expanding the number of subscribers

The simple radiotelephone description suggests that the number of phones and hence subscribers could be expanded quite readily by allocating sufficient radio spectrum to the radio telephone service and therefore having a large number of duplex traffic channels. The shortcomings of this approach are however easily appreciated.

• Firstly, 40 MHz of allocated spectrum really means 20 MHz because of duplex operation. If the equivalent of 20 kHz per channel is achieved, this means that only 1000 users, or less than point one percent of the population in cities of over one million people, would have access to the radio channels.

• Secondly, the near-far problem of radio range is a real problem. How does one manage a subscriber at the edge of the radio range? In such a location his signal would more than likely be suppressed by a user close to his frequency and close to the base station.

• Thirdly, how does one achieve an orderly coverage of a complete state or country and allow a subscriber to roam smoothly through a mobile telephone network only employing the allocated number of radio channels?

1.3   The cellular principle

The required step forward is called the cellular principle. Put simply, radio cells, defined by a base station at their centre, are distributed evenly in clusters with an allowed overlap of coverage. Figure 1.5 illustrates the beginnings of such a distribution of cells, when laid out on a nice flat plain.

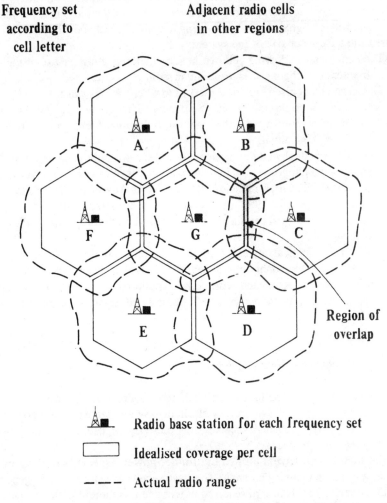

Figure 1.5   The concept of cell distribution and the honeycomb pattern

The first observation, after examining the diagram, is that a different set of frequencies must be used (allocated) to all the adjacent cells around the cell at the centre. However, the same set of frequencies may be used at more distant cells outside the diagram.

This is the principle of the repeat cell and reuse frequencies of operation. The physical distance from the centre of the centre cell to the repeat cell centre is called the *reuse distance*, D. The cell size is defined by its *radius*, R. The relationships between the terms D and R are discussed in Chapter 3.

An important aspect of the reuse of the same radio channels and frequencies in nearby cells means that cellular radio operation to some extent is limited by pre-planned *co-channel interference*, whilst the ability of the particular modulation strategy used for signalling and messaging to combat co-channel interference will determine the cluster size of the system.

The approach of allowing cell overlap and then repeating the use of the same radio frequencies (channels) is the cellular radio principle.

There is no recognised historical mark as to when the idea was first put forward; to some extent one could say that once the concept of the frequency synthesizer was put forward and demonstrated in the early 1960s the opportunity to have frequency active mobiles came about.

The first practical cellular system to appear was the Japanese (AMPS - automatic mobile phone system) in metropolitan Tokyo in 1979, followed closely by the Nordic (NMT450) system in 1981. The North American (AMPS 800) MHz system began service in 1983. (The definition and characteristics of the many national and international cellular systems are given in the Appendix at the end of the text.)

When a mathematical model is applied to the cellular radio cell layout principle a honeycomb of hexagonal cells appears. This is the reason why a hexagonal pattern is generally associated with cellular radio on the covers of conference proceedings, books, etc; however, in practice the real physical radio layout is less well defined.

1.4   Radio coverage by a single cell

A single radio cell and the factors which dictate coverage are illustrated in Figure 1.6. The base station will usually be well sited, have a suitable transmit power, say in excess of 10 W, a sensitive receiver, low noise figure, useful antenna gain and also be clear of site noise.

The mobile will have a limited transmitter power, especially in the portable mode, and an elementary antenna. The base receiver performance is generally unable to make up this loss of received power and it is therefore the reverse path which limits the radio range.

Three ranges are shown in the diagram.

(i)   The operating range - distance: d

(ii) The maximum radio range, i.e. cell size limited by noise, propagation factors
      and transmitter power - called: R_{max}

(iii) The cell size designed for the system, which will be less than R_{max}, decided
      by the coverage and cell pattern considerations - called: R

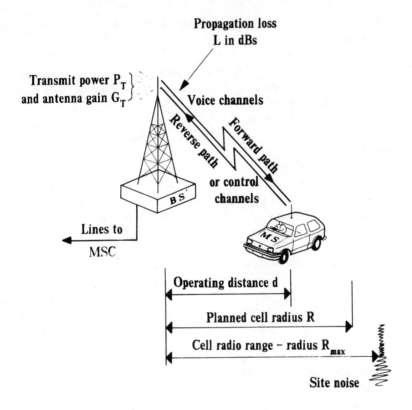

Figure 1.6   The fundamental radio cell and associated parameters

For a flat terrain, R can be regarded as the radius of the cell mapping a circle.
The area covered will thus be πR^2 (km^2). Examples of coverage are shown in
Table 1.1.

Table 1.1   Cell area and number of subscribers covered

Cell radius km^2	Coverage area km^2	Number of subscribers covered*
1	3.14	100
3	28.3	900
10	314	10,000
25	1960	60,000

*   Assumes around 30 subscribers per sq. km as an initial assumption in order to get orders of magnitude.

Adding a column showing the number of users which could be served by such a single cell illustrates how the single cell philosophy of producing radiotelephone coverage will break down or saturate.

The only way forward is to provide lots of small cells - the cellular principle.

Chapter 2 covers the physical principles on which cell radio coverage is established in practice and provides some numerical data for various conditions.

It turns out that for a large cell size, a high, well-mounted base station operating at VHF frequencies is best.

On the other hand, high UHF frequencies and base stations at building height provide compact cells in a city, where the largest density of users is to be found, and hence many cells may well be required.

Table 1.2   Classes and transmitter power of mobiles

Equipment type	Mobile Tx power	Antenna arrangement
IV Handportable*	< 1W	on set and shortened
II Transportable*	< 5W	on set
I Vehicle mounted	> 10W	mounted on vehicle

*   Note: These can sometimes be plugged into a vehicle-mounted system.

The mobile depicted in Figure 1.6 shows a vehicle to emphasise mobility of the end user or subscriber. In practice three main types of mobile station can be defined as shown in Table 1.2.

Clearly the radio coverage or cell size planning will need to take note of the type of user service anticipated, since a vehicle system may well have a 20 dB (or up to twice distance) advantage over a hand portable, because of more favourable transmitter power and antenna positioning.

1.5   Multiple cell layout

Operation in a single cell is the domain of the well established and understood *private mobile radio* (PMR). The use of a *repeater* (central transmitter) can be likened to a central, well-placed base station and large areas can be covered by a single cell or repeater, i.e. 25 km radius or more at VHF. For greater coverage on-frequency (FM) or off-set frequency (AM) technology is or has been used. This is especially important for say police operations in rural areas. For the lower power public radiotelephone concept of service, cell sizes will be smaller, but the same cell frequencies cannot be used again in the adjacent areas.

Figure 1.5 illustrated this and the matter was also discussed, but three further factors now need attention:

• Can one have other frequencies in the adjacent cells?

• How does one manage the region of overlap?

• When can the same frequencies be used again?

The first point emphasises that a cellular radio system will need a block allocation of frequencies. Whether the voice and signalling (control) are managed on a time division, a frequency division, or some hybrid arrangement does not matter, what is important is that much more bandwidth than the bandwidth required by the subscribers in the area of a single cell must be allocated.

This point also emphasises that a mechanism must exist for handing over, called *handover,* the subscriber's radio connection from the frequencies used in one cell, say G, to the frequencies used in the next cell C, as he or she roams the service area of the cells.

This is the feature essential to any cellular radio operation, whether one is considering very small cell sizes (micro-cells), or average, or large cells - sometimes called macro-cells. Much of the complexity within a cellular radio unit, depicted in Figure 1.3, the frequency synthesizer, control and memory functions, are there mainly to handle handover. The radiotelephone circuit system must also be frequency agile to operate in a cellular system. However, other benefits also arise from this attribute, as described later.

The second point is the question of the physical overlap of cell coverage - not a feature apparent in drawings of cellular radio and many other artistic renderings of architectural and artifact aspects with a cellular appearance. The mobile will not know that it is in such a region unless it undertakes a scan of the spectrum. Usually only the base station will know and clearly base stations must be able to communicate with each other. There is therefore a need for an overlay network of communication associated with each BS in Figure 1.5. This network, which can be regarded as a fixed network, can be realised by cable, microwave link or even satellite link.

With the existence of such a connection, station G can communicate to station C say, and they can decide between themselves when to take command of the subscriber moving through the region of overlap; the process of handover. The practical strategy used will depend on the system and other factors and is a fairly complex signal control mechanism.

The fact that several channels or frequencies are used in just one cell in order to accommodate many subscribers, plus the fact that further sets of frequencies must be used in adjacent cells, brings us to the third point; namely, when can the same frequencies be used again? This is the fundamental feature of cell layout strategy and is described in Chapter 3. As an example of a cell layout plan, Figure 1.7 shows the plan for London, UK, at an early stage of implementation.

The layout plan illustrates three useful features to the present discussion.

- Cell sizes are made smaller at the centre of a city or area of occupation of most subscribers.

- Cells are arranged in clusters. Only certain cluster sizes are possible, due to the geometry of a hexagon, and the allowable cluster sizes of 4, 7 and 12 are shown by way of illustration.

- Cell splitting is a permissible operation, that is, installing additional base stations, within, or at the corners of a cell, when an increased density of coverage is warranted.

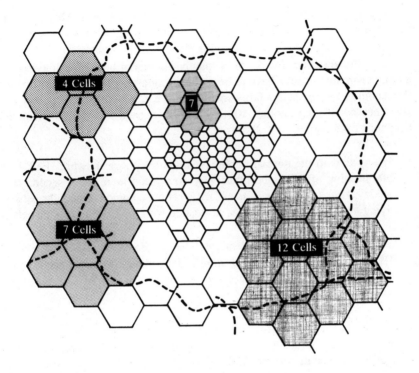

Figure 1.7   A typical city cellular radio cell plan; the cells are smaller where the most users are expected. Cluster sizes of 4, 7 and 12 cells are also indicated

1.6 The fixed supporting network

The fixed supporting network associated with a cellular radio scheme, an outline of which is sketched in Figure 1.8, which is a sophisticated network, has several fundamental tasks:

- To connect all the base stations to each other for the purpose of communicating signals and messages to and from subscribers operating in the network. The modern trend is to use *base station controllers* (BSC) controlling a BS cluster. (In analog cellular the controller forms part of the mobile switching centres, for example, see Figure 3.20.)

- To provide switching centres so that traffic can be directed around the network. The centres are called *mobile switching centres* (MSC). An MSC does not have to be associated with every cluster of cells, such as the seven shown in Figure 1.5, but they are usually sited at convenient town centres for example.

As a cellular radio network grows, and the number of subscribers increases, the mobile switching centres have to begin handling a very large amount of traffic. What could have begun as a fairly small switch, e.g. 100×100 crosspoints, that is 100 lines connecting another 100 lines, must expand to a full capacity fixed telephone exchange.

- Again, as with the normal public switched telephone network (PSTN), full intermeshing of MSCs becomes very costly and therefore a second tier of overlay *transit switching centres* (TSC) is now common in cellular networks, giving a hierarchical network as associated with the fixed telephone network.

- With each mobile switching centre will be associated a *home location register* (HLR) and a *visitors located register* (VLR), although the registers themselves need not be physically associated with the location of their MSC, since the fixed network gives full connectivity.

- Also, *authentication* is provided for a subscriber attempting to use the network. Authentication is usually associated by the home location registers, but is shown as a separate centre (AUC). All equipment used on the network, whether a hand portable, or vehicle-mounted phone carries an electronic identity number, the mobile identity number (MIN). This is usually a ten-digit number programmed into the mobile on registration by a service provider. Meanwhile an *equipment identity register* (EIR) checks out the status of the subscriber identity number. These centres are all shown in Figure 1.8.

- The time of use of the network by a subscriber is also recorded in a *billing centre*. In general the charge made is independent of the physical distance separating two users of the same network, since even if the callers are in adjacent cells, or in the same cell, the full cellular fixed network is really put into use. Clearly, some regional strategy can be put into operation, but this consideration has no bearing on the principles of cellular radio.

- Finally, the fixed network, outlined in Figure 1.8, will have (if agreed) a trunk connection into the existing fixed network. This gives the connectivity between the fixed subscribers and mobile subscribers. The fixed subscriber will use a mobile identification number to call; the mobile subscriber will use the usual area code number. In both cases billing will be made at the mobile tariff rate.

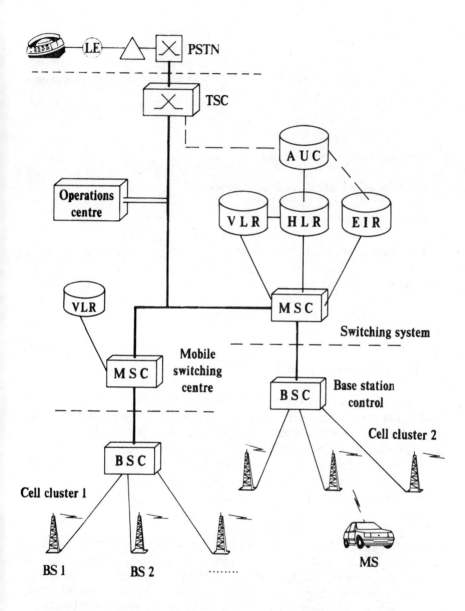

Figure 1.8   A fixed network supporting a cellular radio system showing the various major components

In practice several more mobile switching centres and base stations will exist, but all the main components referred to above are shown. In addition a *network management centre,* nowadays referred to as the *operations and maintenance centre* (OMC), is indicated. There will only be one such centre for each network in general. Its role, as the name implies, includes the following network management functions:

Fault conditions are recognised

Fault diversion strategies can be implemented

Extra traffic routing is supervised

Interference conditions are recorded

Maintenance programs are run

Subscriber base and income are monitored

1.7   Radio frequencies available

We have so far been able to discuss cellular radio without referring to a specific frequency or specific band of frequencies. Not many years ago cellular radio was synonymous with the frequencies 800-900 MHz. Any reference to these UHF frequencies seemed to imply cellular radio. Nowadays, it is recognised that cellular does not have to be associated with a specific band of frequencies; in fact cellular networks exist which range from VHF to UHF frequencies.

There are constraints, however, on those frequencies that can be assigned; these are:

•   The international agreements to use certain bands of frequency for certain users, e.g. broadcasting, mobile radio, navigation, etc. Such decisions are made at World Administrative Radio Conferences (WARC), now held at fairly frequent intervals (in years). In practice, for the allocation of frequencies the World has been divided into three regions. These are shown in Figure 1.9. Often, for historic or geographical reasons, the allocation of frequencies in the three regions has varied.

This could well mean that a cellular service in one region could not be applied directly in another region as the allocation for frequencies is outside the plans of the other region. Some cross border conflict between countries edging regions 1 and 3 could arise for example, but if one takes the Gobi desert straddling these regions as an example, the problem perhaps is not too critical!

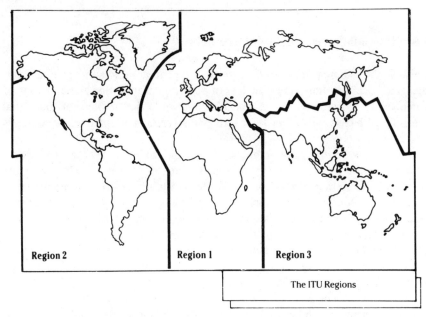

Region 2   Region 1   Region 3

The ITU Regions

Figure 1.9   The three regions in the world according to the International Telecommunications Union (ITU)

- More serious are the National Frequency Allocations. These are the allocation of frequencies by the government or body vested with the right to allocate frequencies on a country by country basis. The frequencies so allocated are listed in the national radio frequency allocation tables, that is, a book listing such frequencies and their uses. These generally differ considerably.

What we shall find is that several cellular radiotelephone systems are in existence (as set out in the Appendix), not from principle, but from frequency allocation. These differences exist especially in so-called *analog cellular,* where digital signalling using frequency shift keying modulation is coupled with analog voice using frequency modulation, and which operate on a frequency division multiple access (FDMA) strategy.

On the other hand, the attraction of using fully *digital cellular,* where both the signalling and encoded speech are digital and more closely attuned to the digital operated fixed network structure, has necessitated such massive research and development, that single regional cellular networks are being accepted in regionally agreed frequency bands. These new systems are also planned on a *time division multiple access* (TDMA) format, but a complete switch to TDMA is not practical, due to the vagaries of radio propagation in cells at high radio frequencies.

1.8   The radio carrier and some attributes

Because the radio carrier is so fundamental to cellular radio some general comments are useful at this point. The loss of power on propagation and operation of antennas are considered in the next chapter, while the problem of the signal reaching both base and mobile by several paths - called multipath propagation - is described in Chapter 5.

A carrier wave of a signal frequency f_c is shown in Figure 1.10(a). It reaches a peak of amplitude twice every cycle, and also crosses zero at twice the cycle rate $t_c = 1/f_c$. An important projection is as a phasor, Figure 1.10(b). The amplitude of a phasor has a constant (peak) amplitude. Its direction defines the carrier phase with respect to a particular time datum. This phasor is projected further to show a progressive (radio) wave, depicted Figure 1.10(c). This latter diagram will be recalled in the next chapter. Also shown is a second possible carrier wave of the same frequency, but in phase quadrature. This wave travels forward with the first wave, but always remains 90° out of phase.

The projection here is again imaginary, however, because it really represents two components which are each a carrier wave, but now drawn as one on paper rotated through 90°. If we were to project the second carrier back to (a) we would obtain a single carrier wave, but whose sent amplitude a was $\sqrt{2}$ times larger.

If on the other hand the two carrier waves were of a slightly different frequency, that is $f_1 - f_2 = \Delta f$, the phasor of f_2 would be rotating at a slightly different speed compared to f_1, and therefore would appear to have a progressively changing phase (and zero crossing times). The appearance of projection (c) could be kept very similar, but the appearance of projection (a) would be that of two waves beating. The beat frequency is the difference frequency Δf. This concept is important to us when multipath propagation is discussed, since it is then easier to understand signal fading and other related properties.

The concept is also very important to us when we consider modulation, in Chapter 6. In analog cellular, frequency modulation (FM) is used. Clearly, what one does here is to rotate the phase back and forth in sympathy with the voice waveform signal. The mobile, or base, receiver demodulator has therefore to follow the phasor around the phase circle faithfully.

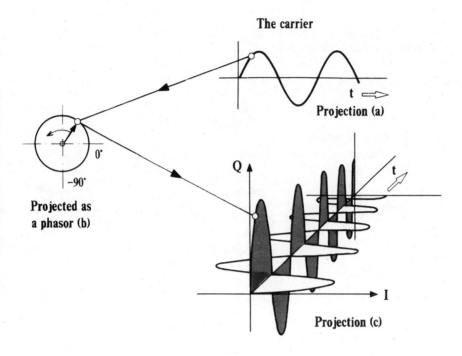

Figure 1.10   A carrier wave depicted as (a) a conventional waveform, (b) projected as a phasor, and (c) projected into phase space and two carriers progressing.

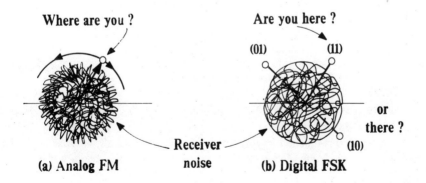

Figure 1.11   A frequency modulated carrier vector subject to noise, (a) analog FM case (b) digital FSK case

This works well until the noise level in the system (site plus receiver circuit's) becomes excessive, as suggested in Figure 1.11(a); just beyond this threshold, the noise phasors take over and the receiver output (radiotelephone arrangement) becomes noisy and unusable. This limit is set by the carrier-to-noise ratio (C/N), a power ratio in dB, at the receiver. The effect is demonstrated in Figure 1.12. This defines the distance R_{max} in a cell site.

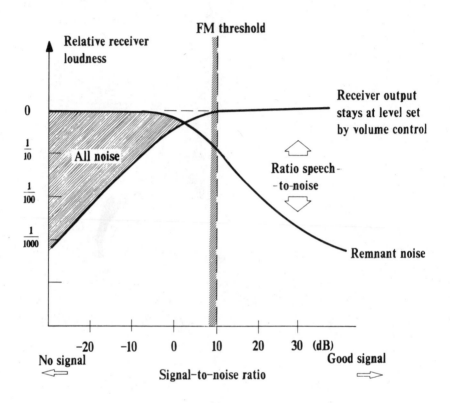

Figure 1.12   The capture effect in FM demodulation indicating acceptable signal-to-noise ratios at a receiver

Clearly, it would be undesirable to approach this FM threshold point too closely; therefore handover is arranged by observing, at the base site when a somewhat higher C/N ratio is reached, say +20 dB. This defines the cell radius R. Note that one has therefore implied that the C/N ratio is used to effect handover in analog cellular.

The situation is different in digital cellular. The modulation will be a digital modulation which can in general be regarded as being a frequency shift keying (FSK) format, but at any specific digit symbol time-of-occurrence the carrier will find itself in some specific phase location space, as depicted in Figure 1.11(b). Noise or interference in this case is shown as confusing the wanted signal phasor. The receiver demodulator does not now produce additional noise, but the decoded data now contains incorrect digits, giving rise to what are called bit errors. These are measured as bit error rate (BER); for example one error in 100 bits, BER = 1×10^{-2}, in 1000 bits, the BER = 1×10^{-3}, etc. In fixed computer networks one assumes really no errors or perhaps a BER of less than 1×10^{-11}.

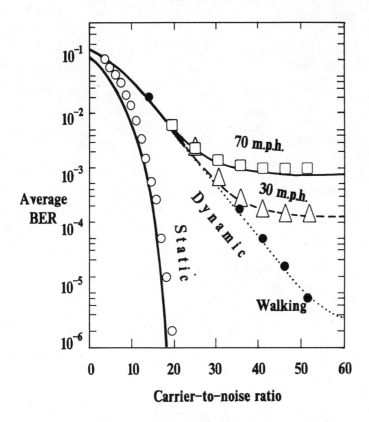

Figure 1.13   The dependence of bit error rate (BER) on carrier-to-noise ratio and the operating conditions of the mobile receiver

In mobile radio such a low error rate is extremely unlikely. A more likely performance is suggested in Figure 1.13. This diagram brings out several points which we need note.

• The static case refers to the particular digital signalling technology chosen, as discussed in Chapter 6.

- The dynamic case refers to the mobile situation. The faster the subscriber travels the more serious the BER problem becomes. This is due to a relationship between the multipath signal characteristics and the travelling speed of the mobile.

- In certain situations the BER can reach an irreducible level. Increased radio power within the cell is of no avail. Emphasis must therefore be put on *forward error correcting* (FEC) strategies. FEC thus becomes a major aspect of signalling within a cellular radio system.

1.9   Control and channel signalling

A mobile radiotelephone system differs from a fixed telephone network in the important operational aspect of roaming, or rather the subscriber being able to appear anywhere within the network at anytime. Whereas many years ago all telephones tended to be black and have a sameness, mobility brought with it a need for a more colourful system and active system management. To some extent we have an analogy with the historic system of signalling in the Navy; a ship would show its colours in order to be identified, or to pass messages. So in cellular radio coloured messages are exchanged between subscribers and their mobile phone sailing through the system.

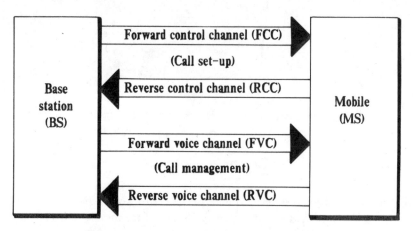

(This scheme principally applies to AMPS and TACS)

Figure 1.14   The control and signalling channels associated with an analog cellular system

More specifically there are four channels set up between a mobile and its associated base as shown in Figure 1.14 and Figure 1.6. In FDMA systems, such as TACS and AMPS these would be four radio channels. In a digital TDMA system the channels would be allocated time slots; pairs allocated to control, other pairs allocated to the phone conversation and its management. Other variations are possible, such as in the Nordic Mobile telephone (NMT) system where channels have reversible roles.

In general control channels (or time slots) are used to page or call the mobile. A response from the mobile is required by the system in order to check its identity, *authenticity* (friend or foe) and whether it is within satisfactory communications range. Thus cellular radio depends very much on a handshaking procedure. In addition the cells are colour coded by a simple two-bit symbol, called the *digital colour code* (DCC), assigned to the control channels; this assists the system to manage frequency (or time slot) assignment in a cluster of cells. Hence the analogy with naval operations. When agreement has been reached between the mobile and the system, which takes place by means of a medium speed digital signal using a *high-level data link control* (HDLC) format, the system and mobile set up the communication circuit on the *voice* channels, as indicated in Figure 1.14. Here again *supervisory audio tone* (SAT) signals are an important part of the final handshaking or call set-up operation. Cell clusters are now SAT coloured by three tones. An illustration of the whereabouts of the DCC coding and SAT allocations in a cellular radio cell coverage group is given in Figure 1.15.

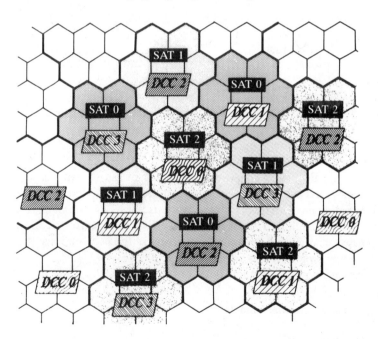

Figure 1.15   Cell layout plan showing allocation of DCC codes and SAT tones

For the majority of control interchanges analog cellular uses high speed digital signalling with high redundancy to give high reliability even in poor radio conditions. Each signalling word is repeated several times, and at the receive end a bit-by-bit majority decision is performed which corrects the burst and random errors. The signalling word is further protected by a Bose - Chaudhuri - Hocquenghem (BCH) block code, which is used to check for residual errors, and allows for up to one remaining bit error to be corrected. A typical frame is shown in Figure 1.16.

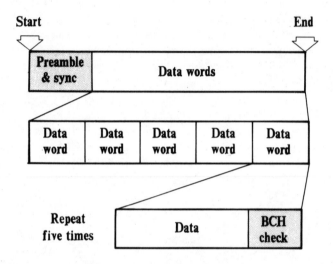

Figure 1.16   Signalling message format on the control channel

On the base-to-mobile signalling link or voice channels, where it is particularly important that handover commands are correctly received, often under poor and rapidly degrading radio conditions, signalling messages carry an even higher redundancy, and are repeated up to eleven times, as indicated in Figure 1.17.

Depending on the network, the network uses a set of control channels which carry only signalling. The channel numbers of the control channels are pre-designated, and all mobile stations operating in the network have the numbers permanently programmed in their memory.

Functionally, there are three types of control channel:

Dedicated control channels

Paging channels

Access channels

Physically, however, all three functions may be combined on each of the control channels, and in practice this is normally the case. Most standards allow for the separation of paging and access functions into separate physical channels, and for the extension of the total number of control channels in cases where the amount of signalling activity is greater than can be handled by the combined control channels.

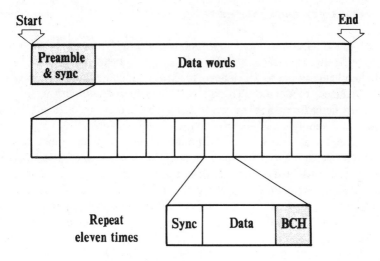

Figure 1.17   Signalling message format on voice channels

The functions carried by the various control channels may be summarised as follows:

- Dedicated control channels - unidirectional (base-to-mobile) - used to carry basic system parameters and information on the channel numbers of the paging channels currently in use.

- Paging channels - unidirectional (base-to-mobile) - used to carry basic system parameters, location area identity, and paging messages to specific mobiles for incoming call set-up.

- Access channels - bidirectional - in the mobile-to-base direction, used to pass all information to the base station concerning the call set-up request or location registration by the mobile; in the base-to-mobile direction, used to confirm location registrations and to assign a voice channel following a call set-up request.

The signalling within the fixed network of a cellular system is identical to that used within all present-day digital fixed telephone networks and runs according to

CCITT (telephone and telegraph committee of the ITU) standards. Because digital cellular technology is much more closed tuned to the modern digital fixed telephone technology, a fuller description of the fixed network in support of mobile phones is deferred until Chapter 8.

1.10 Error correction strategies

It is widely recognised that in practice mobile radio signals in any radio band, but more specifically at VHF and UHF, are always subject to variable received amplitude, changing phase and indeed spectrum widening due to a propagation induced frequency modulation. This is the phenomenon of *multipath reception*; the signal components from one base transmitter appear to come by different paths and directions to the mobile and give rise to these multipath effects.

Cellular radio can be designed in principle without giving much regard to multipath propagation. Narrowband analog FM voice transmission performs acceptably well, provided that careful voice bandwidth shaping is introduced in the transmitting and receiving *baseband* circuits. The problem comes with the control signals used to set up and manage the system, i.e. the signalling procedures just described above. Multipath now causes real difficulty due to error generation and false signalling. The particulars of multipath are described in more detail in Chapter 5 and in particular the concept of *coherence bandwidth* and its inverse, the *delay spread* are introduced. When the bit rate used for signalling becomes comparable with the delay spread (measured in micro seconds), error correction strategies become vital.

There are two basic approaches to error correction. One is the *automatic repeat request* (ARQ) arrangement outlined in Figure 1.18. This method is used in a data specific cellular system described later in Chapter 4.

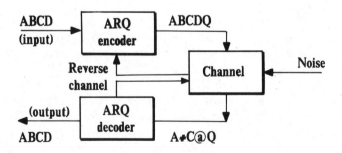

Figure 1.18   Automatic repeat request system

On the other hand *forward error correction* (FEC) aims to correct the message once and for all. Like ARQ, forward error correction has to be implemented at both the mobile and the base, so there is a penalty on hardware and power consumption in the mobile, expecially if it is handportable. FEC implementation is indicated in Figure 1.19.

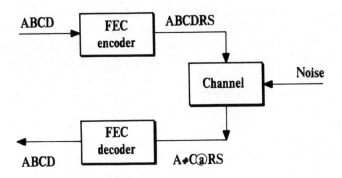

Figure 1.19   Forward error correction implementation in a system

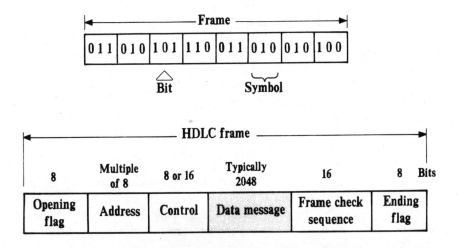

Figure 1.20   The construction of an HDLC frame and nature of bits and symbols

In order to discuss the FEC options we need to define the type of signal message likely to be encountered. The smallest unit of data is the *bit;* a *symbol* may be formed by several bits if a multi-level modulation scheme is used. A group of symbols are then made up into a frame. Figure 1.20 shows a frame made up of 9 symbols with three bits per symbol; however, only some are *message* symbols, several could be redundant symbols introduced for the purpose of forward error correction.

In TDMA systems, as described in Chapter 8, a single frame as shown here is now deemed a *time slot* (TS) within a full frame made up of several time slots - all carrying different data.

The encoder takes a set of k message symbols which must be transmitted (pure data), appends to them r check symbols, and transmits the entire block of n = k + r channel symbols, the frame. The result is denoted an (n,k) code. If the channel noise distorts sufficiently few of these n transmitted symbols, the r check symbols will provide the receiver with sufficient information to enable it to detect and correct the channel errors. Since each codeword contains n symbols and conveys k symbols of information, the theoretical information rate is defined as $R = k/n$.

There are two major classes of error-correcting codes, *block codes* and *convolutional codes,* and the essential difference between them is as follows. In a block code, the n-symbol codeword generated by the encoder in a particular time unit depends only on the k message-symbols received within that time unit. In a convolutional code, the frame of n code-symbols generated in that time unit depends not only on the current message frame, but also on the preceding $N - 1$ message frames, where N is referred to as the constraint length. Hence, as each incoming information bit propagates through the encoder, it influences several outgoing bits, thereby spreading the information content of each message bit among several adjacent code bits. Usually the values of k and n are small; in fact, many convolutional techniques operate with a code efficency of 50%, i.e. one check bit inserted after each information bit.

Under convolutional codes, the Viterbi code corrects random error-patterns, that is, patterns in which any one particular bit has as much chance of being distorted as do the bits immediately preceding or following it. Such an error distribution arises from *additive white Gaussian noise* (AWGN). Block codes, on the other hand, correct burst errors, i.e. where a number of errors occur closely together, separated by a large guard-space (or error-free zone). Such an error distribution arises from impulse noise and there is a dependence of error probabilities in successively transmitted bits.

Under block codes, the Reed-Solomon and Golay codes have both burst and random error-correcting abilities; the former treats non-binary data, whereas the latter handles only binary data. The Reed-Solomon code is a special sub-class of the Bose-Chaudhuri-Hocquenghem (BCH) code, with powerful applications for correcting multiple bursts of errors in a frame. BCH codes themselves are a sub-class of the cyclic codes, which in turn belong to the family of linear block-codes. The Golay code is capable of correcting a combination of 3 or fewer errors in a block of 23 bits and is the only known multiple-error-correcting binary perfect code.

A simplified family tree of codes is given in Figure 1.21, showing the relative positions of the codes within the coding hierarchy.

One might well question how efficient these error-correcting codes are. The information rate has to be reduced, which in turn implies fewer bits per bandwidth of the radio channel allocation.

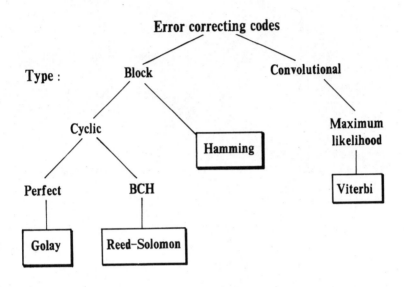

Figure 1.21   An error-correcting code hierarchy diagram

Much study of this matter has taken place, to some extent prompted by deep space satellite experiments, where the received signal is at the edge of detectability. This limit was defined by Shannon, who showed that the theoretical limit of the detectability of an energy per bit E_b, to Gaussian average noise power per unit bandwidth N_0, can be -1.6dB. Practical systems need many decibels of power above this level. Figure 1.13 shows how BER changes versus received carrier-to-noise power ratio for a typical mobile radio system. Translated into the BER versus the ratio E_b/N_0 (described further in Chapter 6), shows that any particular modulation, such as differential phase shift keying (DPSK), needs many decibels of excess received power as compared to the Shannon limit, i.e. Figure 1.22.

However, adding the FEC codes described above, shows how much improvement can be gained, especially if initially an error rate is less than one bit wrong in, say, 200 bits. The error rate can be improved to 1 in 10,000. However, only when the pure data channel is relatively free of errors, can any significant reduction in the ratio E_b/N_0 be made. In all cases, however, FEC introduces the penalty of a slower throughput of data.

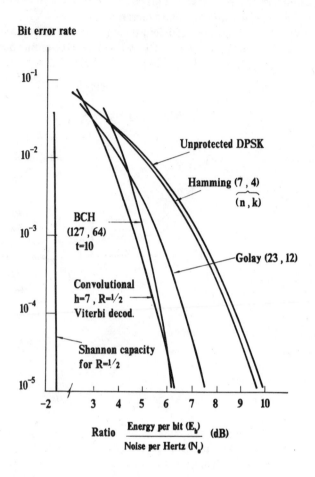

Figure 1.22   Improvement of BER performance by specific error-correcting
                codes and the Shannon limit

1.11 Numbering plans

Although we have referred to numbering plans earlier in this introductory chapter,
the number plans associated with the fixed network (PSTN) and the mobile
equipment identification is quite complicated and expansion of some points is
worthwhile at this stage.

The original numbering plan for the international telephone network is described in CCITT (the International Telegraph and Telephone Consultative Committee) Recommendation E.163. In this scheme each country or zone is assigned

a country code of 1, 2 or 3 digits, with a maximum international number length of 12 digits (excluding prefixes such as 0, 00 and 010, etc). This scheme served until the early 1980s when consideration was given to the numbering aspects of the *integrated services digital network* (ISDN). The numbering plan for the ISDN needs to evolve from the PSTN, so the numbering plan for ISDN is an evolution of the existing E.163 Recommendation.

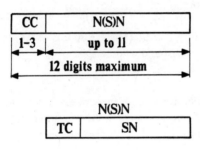

CC   : Country code
N(S)N : National (significant) number
TC   : Trunk code
SN   : Subscriber number

Figure 1.23   E.163 number structure

This is known as Recommendation E.164 'The numbering plan for the ISDN era' and was approved in 1984. The main principles of E.163 remain, but with the introduction of an increased international number length to 15 digits, and the introduction of a *network destination code* (NDC) in place of the E.163 trunk code. E.163 allows for up to 2 digits to determine the international route; this gave a variable maximum number analysis of 3 to 5 digits, depending on the length of the country code; in E.164 this was fixed at a maximum of 6 digits including country code. This change reflects the increased capability of *stored program control* (SPC) exchanges.

Recommendation E.165 details the arrangements for the implementation of the E.164 plan and specifies the date of 31 December 1996 for bringing the recommendation into effect. The present world fixed network numbering zones, as set out in Table 1.3, are to be maintained, however.

Table 1.3   World fixed network numbering zones

Code	Zone
1	North America (including Hawaii and the Caribbean)
2	Africa
3 & 4	Europe
5	South America and Cuba
6	South Pacific (Australasia)
7	USSR
8	North Pacific (Eastern Asia)
9	Far East and Middle East

The identification plan for mobile subscribers is contained in CCITT Recommendation E.212, and the allocation of mobile station roaming numbers is defined in E.213, shown as Figure 1.24. The number consists of:

- *Country code* (CC) of the country in which the mobile station is registered, followed by,

- National (significant) mobile number which consists of *network destination code* (NDC) and *subscriber number* (SN).

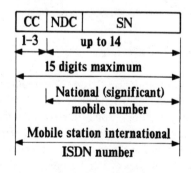

CC   : Country code
NDC : Network destination code
SN   : Subscriber number

Figure 1.24   The E.164 number structure and mobile numbering structure

The number of digits used for mobile identification is limited to a ten-digit code, made up of the CC, the NDC and the SN. These are better known in the cellular radio business as MIN numbers. MIN 1 is the country code, but is now a *mobile country code* (MCC) and will be three digits. Table 1.4 shows examples of MIN 1 or MCC codes. The first digit refers to the world numbering zone; note that *these zones are not the same zones as used by the fixed network.*

Table 1.4   MIN 1 (MCC) numbers

Zone	Country	Country Code
2	Belgium	206
2	Italy	222
2	Malta	278
3	Canada	302
3	Barbados	342
3	Bahamas	364
4	Jordan	416
4	Qatar	427
4	Maldives	472
5	Australia	505
5	Fiji	542
5	Cook Islands	548
6	Egypt	602
6	Tanzania	640
6	Zimbabwe	648
7	Panama	714
7	Bolivia	736
7	Uruguay	748

The MIN 2 number consists of seven digits and usually corresponds to the mobile unit telephone number.

CCITT Recommendation E.213 states the following:

• The numbering plan should allow standard telephone charging and accounting principles to apply.

- Each administration should be able to develop its own numbering plan.

- It should be possible to change the international roaming identity without changing the telephone number allocated to the mobile.

- Roaming without constraints should be possible.

This numbering plan refers only to interconnection with the fixed telephone network (that is, it does not apply to mobiles that are not interconnected).

The MIN number is loaded in the mobile as part of the *number assignment module* (NAM). This number also contains data pertaining to its home location register.

The NAM can be a PROM (Programmable Read-Only Memory), an EPROM (Erasable Programmable Read-Only Memory), or an E2PROM (Electrically Erasable Programmable Read-Only Memory) that contains details about the customer, the system, and the options chosen.

In addition an *electronic serial number* (ESN) is added to the MIN1 code. This number enables the network to recognise that the mobile equipment being used is valid and indeed registered with the network.

1.12 Summary of important features

Although many cellular systems exist and are being implemented in the world today, it can be said that they all depend on certain basic principles, despite being totally incompatible in many ways. Thus, we can define the following list of features that the systems have *in common,* namely:

- At any instant, all subscribers (the mobile) operate in a single numbered radio cell.

- The radio properties of a cell are those of propagation coverage and multipath effects.

- The cell is one of a very large number laid out on a specific plan arrangement, made up of clusters of cells.

- The frequencies assigned to each cell in a cluster are repeatedly used in other clusters.

- The mobile therefore operates expecting both co-channel and adjacent channel interference.

- The mobile must register with a fixed telecommunications network which controls the operation of all the cells.

- The network can assist the mobile to be handed over from one cell to another as it moves geographically. This provides the roaming and a no-fixed-abode facility.

- To manage these last two features the mobile must be constantly signalling to and from the fixed network.

- The message channels are duplex and can be voice or data.

- The fixed network can interconnected to the public telephone network.

- The mobile telephone must be frequency agile and carry batteries and be given an identity number.

Differences between cellular radio systems come about because of the following features:

- The radio frequency band(s) allocated to the specific service.

- The mode-of-operation can be all-digital or analog plus digital.

- All-digital systems allow time division access as the method of having many subscribers within one cell.

- Mixed systems use frequency division access methods for sorting out subscribers.

- There are variations of the basic parameters of the FDMA and TDMA systems.

- The above mode-of-operation differences give rise to different handover strategies and details of the fixed network architecture.

- The more recent systems have only been made possible by developments in VLSI circuit technology.

In all cases the density of radio cells within a geographical area, and the effective number of radio channels available, determine the capacity of a system, and this is discussed in the last chapter.

Further reading

A Guide to TACS - Total Access Communication Systems. (1985). DTI publication UK

Boucher, J.R. (1990). *Cellular Radio Handbook,* Quantum Publishing Inc, USA

Clark, G.C and Cain, J.B (1981). *Error-correction Coding for Digital Communications,* Plenum Press, NY

Ericsson Review, (1987). 'Special issue on cellular radio telephony', No 64

Farrell, P.G. (1990). 'Coding as a cure for communication calamities', *Elec Comms Eng J.,* December, pp 213-220

Jakes, W.C. (1974). *Microwave Mobile Communications,* Wiley-Interscience, NY

Lee, W.C.Y. (1989). *Mobile Cellular Communications Systems,* McGraw-Hill Book Co, NY

Personal Communications IC Handbook (1990), GEC-Plessey Semiconductors Ltd, UK5

Philips Telecommunication Review. (1983). 'Special issue on mobile radio', April

Young, W. R. (1979). 'Advance mobile phone service: Introduction, background and objective', *B.S.T.J.,* 58, pp 1-41

2   Radio Coverage Prediction

2.1   Electromagnetic waves

Radio waves form part of the naturally occurring electromagnetic wave spectrum, any one frequency component of which can be represented as a progressive electromagnetic wave. Such a wave component is depicted in Figure 2.1.

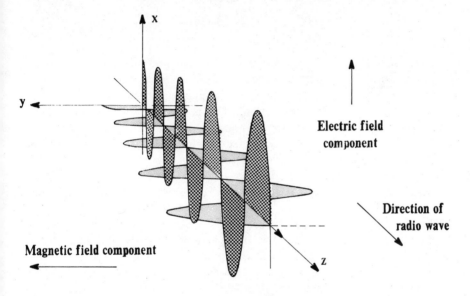

Figure 2.1   Representation of an electromagnetic wave moving through space showing the magnetic and electric oscillatory components

It is important to understand this diagram. Firstly, the wave travels forward in direction and this is the basis of the concept of directivity of an antenna. The waveform repeats itself after a distance of one wavelength. For mobile radio this distance or wavelength ranges from metres to centimetres. Although the frequency of the wave, say 100 to 1000 MHz, is specified in the assignment of a channel to a mobile radio user, wavelength is more relevant to the radiation and reception of the EM wave when considering the antenna. Also the defects of propagation of the wave, such as destructive multipath, when replicas of the same wave arrive from different directions, takes more note of the wavelength.

Secondly, Figure 2.1 indicates that the EM wave has two active components, the electric field vector and the magnetic vector. These are in phase in time but not in space. The diagram shows a *vertically polarized* radio wave. This polarization

37

arose because of the way it was generated. Because there is a tendency to use vertically mounted dipoles or monopole antennas, vertical polarization ensues, with a vertical electrical field component, usually measured in microvolts per metre, or dBμV/m.

Unless one is able to markedly alter the electrical characteristics of free space, the magnetic component of the wave is closely related to the electric component value, from which one obtains the radiation power density (S) of the EM wave. This is measured as the electrical power (watts) passing through a plane one meter square, facing the direction of the wave.

The relation is given by the well known equation

$$S_R = \frac{E_R{}^2}{120\pi} \qquad \qquad \dots (2.1)$$

120π is called the intrinsic impedance of the free space, as can be seen from the relation between the voltage and power, i.e. approximately 377 ohms.

A third attribute of Figure 2.1 is that if the page of the book is rotated the effect of a change of polarization can be noted. Ninety degrees gives a horizontally polarized wave, which will not be detected by a vertically orientated electric antenna (dipole). It is to be noted that a change of polarization can occur at sharp angles of ground reflection.

A circularly polarized wave is one which has a continuous rotation of polarization induced into it by the radiating antenna structure. Current mobile radio systems use vertically polarized signals, but it may be found advantageous to use more complicated wave structures at new bands in excess of 1 GHz; indeed the signal is most likely to be very complicated as shown in Chapter 5.

A final feature of Figure 2.1 is that the diagram only shows a single EM carrier wave, that is, a single frequency. An actual mobile radio signal will consist of a closely knit group of such waves, being a modulated carrier wave. The forward movement of the signal is not generally taken into account, but certainly manifests itself under mobile conditions as described in Chapter 5 on multipath propagation.

2.2   Antenna considerations

In order to describe radio wave propagation it is necessary to have a basic appreciation of antenna theory, but the requirements can be kept very brief. Thus the generation of the wave depicted in Figure 2.1 required, for example, a conducting structure in which there is an oscillating electrical current.

The simplest antenna is known as a *dipole,* formed by separating the ends of a transmission line or coaxial cable. The structure is resonant, or rather matches to the transmission line, when the dipole has a total length of one half a wavelength, hence the name, *half wave dipole,* as indicated in Figure 2.2.

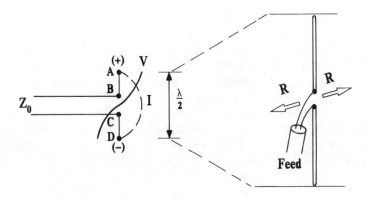

Figure 2.2   Open transmission line forming a half wave dipole

If the top section is mounted above a conducting surface, a monopole antenna
is formed. This need only have one quarter of a wavelength in physical height. Thus
at 900 MHz this implies a short metal rod of about 8.3 cm. Figure 2.3 shows the
reflection at the base of the antenna (the centre of the equivalent dipole).

The roof mounted monopole (half a dipole) is the best known example. The
current oscillates in amplitude and absolute magnitude along the vertical structure
and sends out radiation, mainly in the horizontal plane in which the antenna is
situated. The fact that the radiation pattern is not isotropic implies directional gain
(G). Gain and direction are related because antenna gain means a concentration
of radiation in a particular direction. A short dipole has a gain of about 1.5 times
(= 1.76 dB) compared to an isotropic or point source antenna.

Because simple dipole UHF antennas are physically small, they can be regarded
as a point source when viewed at a distance from the base station or mobile.

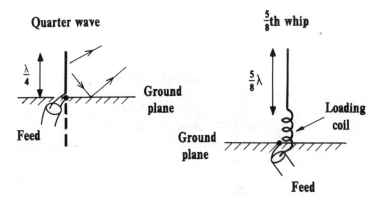

Figure 2.3   The roof mounted $\lambda/4$ monopole antenna. Extra effective height and
                gain can be obtained by a loading coil

For a mobile the antenna can be reduced in length (or size) by what is known as *coil loading*. Sometimes the base loading coil can be seen, i.e. as on a vehicle; alternatively a *helix* is encapsulated in a protected stub which acts as the antenna. Although the antenna remains resonant at the frequency band of operation its effectiveness as a receiving antenna is proportional to the physical height of the antenna. This fact needs to be observed when planning cell sizes.

At base stations the antenna can be more robust, as this gives system gain. Typically a stack of *folded dipoles* is used. Four are shown (vertically) in Figure 2.4. This arrangement will double the gain, i.e. add +3 dB, and double the bandwidth. The folded dipole responds more effectively over a wider band of frequencies.

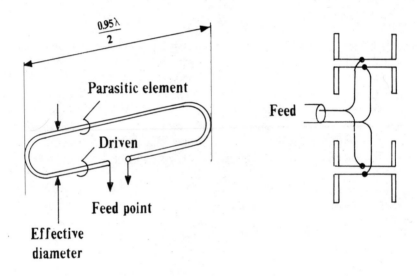

Figure 2.4   A folded dipole, also shown arranged as a stack of four

To obtain a sectored cell pattern a reflecting or director element would be associated with each active folded dipole. The directivity associated with the familiar TV Yagi antenna would be too great, say, less than a 30° beamwidth.

When studying propagation and the associated antennas it becomes helpful to think of an antenna as a collector of radiation, having a certain aperture. This aperture is called the *effective aperture* (A) of the antenna and is the planar size of the antenna, so far as a collector of radiation is concerned.

An antenna of aperture A collects a power P (watts)

$$P = A.S_R$$

$$...(2.2)$$

It can be shown that the gain G and aperture A of all antennas are related by the formula

$$G = \frac{4\pi}{\lambda^2} A \qquad \qquad \ldots (2.3)$$

The formula is most easily understood by considering a microwave dish antenna. The aperture is approximately the dish area (allowing for some loss of efficiency); therefore as the size (diameter) increases, so does the gain.

For a short dipole,

$$A_{dipole} = G \frac{\lambda^2}{4\pi} = 1.5 \frac{\lambda^2}{4\pi} = \frac{3\lambda^2}{8\pi} \qquad \qquad \ldots (2.4)$$

This equation again emphasises how small an aperture handportable equipment UHF antennas present to the radio signal in a radio cell.

Using the previous equations one can calculate the relation between the power received by a dipole P_d and the signal field strength (E) at the dipole. At 900 MHz a useful relationship is

$$P_d \text{ (dBm)} = -135 + E \text{ (dB}\mu\text{V/m)} \qquad \qquad \ldots (2.5)$$

For example, with 10 μV/m, $P_d = -115$ dBm, which is a typical receiver final sensitivity. Lowering the frequency of operation to, say 200 MHz, increases P_d to -103 dBm, as the aperture of a dipole tuned to 200 MHz is larger, because the wavelength is larger.

The signal power received is the critical factor in propagation, because it has to overcome the noise power N_R at the receiver input. With N_R in dBm, the received signal/noise ratio, expressed in decibels, will be

$$(S/N)_R = P_d - N_R \quad \text{(dB)} \qquad \qquad \ldots (2.6)$$

The S/N ratio can be improved by having antenna gain. The result, with a receiving antenna of gain G_R, will be

$$(S/N)_R = P_R + G_R - N_R \text{ (dB)} \qquad \qquad \ldots (2.7)$$

2.3 Models for propagation

Models for radio propagation all begin with the concept of two-point source antennas in free space separated by a distance d (km).

2.3.1   In free space

The radiation power density at the receiver of the signal from the transmitter of source power P_T will be

$$S_R = P_T/4\pi d^2 \quad \text{(watts/m}^2\text{)} \qquad \ldots (2.8)$$

since the radiation is distributed over a sphere of radius d. If the transmitter antenna has a directivity gain G_T, in the direction of the receiver, the received signal increases to

$$S_R = G_T P_T/4\pi \ d^2 \qquad \ldots (2.9)$$

Note that $G_T P_T$ is the *effective intrinsic radiated power* (EIRP) of the transmitter. Furthermore, if the receiving antenna has an aperture A, the power received, according to (2.2), is

$$P_R = S_R.A$$

therefore

$$P_R = S_R .G_R \ (\lambda^2/4\pi) \qquad \ldots (2.10)$$

Introducing the value for S_R from (2.9) gives

$$P_R = (P_T/4\pi d^2) . G_T . (\lambda^2/4\pi) . G_R$$

or

$$P_R/P_T = G_T.G_R \ (\lambda/4\pi d)^2 \qquad \ldots (2.11)$$

Note that directivity gain at either end of the system enhances the received signal, but there remains the fundamental free space propagation factor $(4\pi d/\lambda)^2$ which reduces the received power as compared to the transmitted power. This is called the *free space propagation loss factor* L, and is given by

$$L = (4\pi d/\lambda)^2 \qquad \ldots (2.12)$$

L arises due to dispersion of the wave energy. L increases as the wavelength reduces, i.e. at higher frequencies. This equation is more often written in logarithmic form

$$L_{dB} = 32 + 20 \log f_{MHz} + 20 \log d_{km} \qquad \ldots (2.13)$$

Example   At 450 MHz over a path length of 1 km, the formula gives

$$L = 86 \text{ dB}$$

How does this relate to the transmitter power required to achieve effective communication? One needs knowledge of the receiver noise threshold N_R and the acceptable S/N ratio. Therefore P_T has to overcome N_R and L, but is helped by the antenna gains G_T and G_R.

Therefore

$$P_T > (S/N) + N_R + L - G_T - G_R \quad \text{(dBm)} \qquad \ldots (2.14)$$

Typical observed noise levels are measured in dBm, but are shown in ^0K, using the fact that the effective noise temperature T_a creates the noise N_R, i.e.

$$N_R = k T_a B_w \qquad \ldots (2.15)$$

where

$$k = \text{Boltzmann's constant}$$
$$= 1.38 \times 10^{-23} \text{ Watt sec/}^\circ K$$
$$B_w = \text{system bandwidth in Hertz}$$

Therefore, at

$$T_a = T_0 = 290 \text{ }^\circ K \text{ (+17}^\circ C),$$
$$N_0 = -174 \text{ dBm per Hz}$$

Figure 2.5 shows how much the actual system noise level is above N_0; either as a *noise figure* F_a, or as a *noise temperature* T_a, relative to the reference noise temperature T_0. The data are then independent of the receiver bandwidth. Note that at UHF, receiver circuit noise and external site noise are comparable and together they limit the maximum cell size R_{max}.

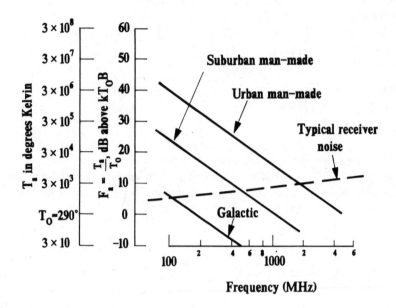

Figure 2.5   Typical noise levels for a UHF receiver

Example

S/N	=	+20 dB
N_R	=	−120 dBm
L	=	86 dB
G_T, G_R	=	3 dB (handportables - helical antennas)

Therefore,

P_T > −8.0 dBm
= 160 microwatts

which is a very low power.

This is the free space result. On the ground, specular reflection modifies the received signal, as shown in Figure 2.6.

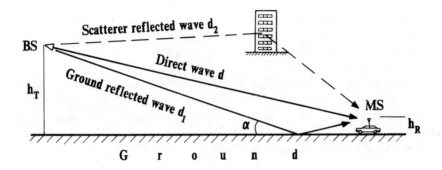

Figure 2.6   The direct wave, ground reflected wave and one scattered wave in the mobile radio environment

2.4   Reflections at a boundary

Two types of reflected wave are indicated: one reflected off the ground (terrain) and onto the mobile; the other reflected or scattered from a surrounding building or hillside. The reflection of an electromagnetic wave at a *boundary*, that is a surface where there is a relative change in the electrical and magnetic properties of the propagating medium for the EM wave, usually from air to a partial conductor, the ground for example, is a complex matter.

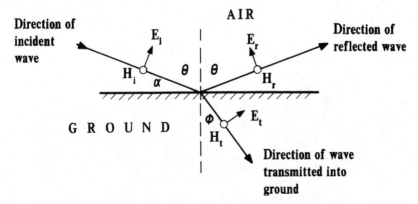

Figure 2.7   The EM wave components transmitted due to a boundary

Figure 2.7 indicates the operation for an incident vertically polarized signal: part of E_i is reflected as E_r, part is transmitted into the ground as E_t. There are also corresponding magnetic vector components of H_i. The complex ratio of E_r/E_i can be calculated, as can the other respective ratios.

What is discovered is that the magnitude of E_r/E_i, called the *reflection coefficient*, is a critical function of the angle of incidence, and also the phase of $E_r : E_i$ is likewise a critical function. Also the reflection coefficient is sensitive to the relative conductivity of the ground.

Figure 2.8 shows results suggested by the CCIR, for frequencies of 1 GHz. The results are plotted against the *grazing angle* α shown in Figures 2.6 and 2.7.

In general one is looking for radio conditions where a truly grazing angle applies.

Four significant observations can be made

(i)   At grazing angles below about 1° there is good reflection when vertically polarized waves are used.

(ii)   At these low angles the electric vector undergoes a phase reversal of 180°.

(iii)   At moderate grazing angles vertically polarized waves are markedly attenuated, but suffer less phase reversal.

(iv)   Horizontally polarized signals do not show the same variation, being generally well reflected, but again with a phase reversal of the reflected component.

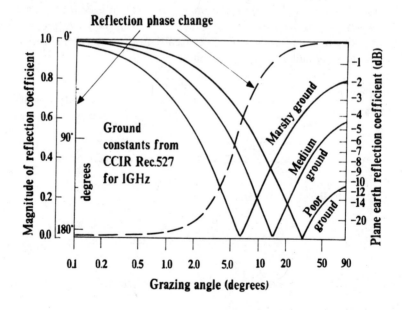

Figure 2.8   Reflection coefficient of a vertically polarized signal for various ground types

In cellular radio one can regard the ground as being horizontal because the distances (cell radii) are generally small, say, in the range of 1-20 km. Thus the grazing angles which apply can be calculated readily from the geometry of Figure 2.6 and the grazing angle is given by

$$\alpha = \tan^{-1} \frac{h_T + h_R}{d} \qquad \ldots (2.16)$$

where h_T and h_R are the effective heights of the BS and MS above the ground plane. Suppose we let $h_T = 30$ m, $h_R = 1$ m, we then find

when d = 300 m $\alpha = 5.9°$
       = 500 m   = 3.5°
       = 1 km   = 1.8°
       = 3 km   = 0.59°

These results indicate that for very small cells or mobile-to-base operating distances, the ground reflection coefficient may be more complex than we assume below, for example.

2.5 Terrestrial propagation

2.5.1 Simple flat earth model

The usual approach, when dealing with the situation in Figure 2.6, is to assume the distance

$$d \gg h_T \text{ and } h_R$$

which gives the relative phase delay between the direct ray and the reflected wave as

$$\phi_d = 2\pi h_T h_R / \lambda d$$

Adding to this the additional phase angle ϕ_r induced by the reflection process, as indicated in Figure 2.8, the electric field received at MS will be the sum of two signals, that is

$$E_S = 2E_R \cos\left(\frac{\phi_r + \phi_d}{2}\right)$$

using the cosine rule, and assuming that the reflection coefficient is unity. If this is the case, and also letting $\phi_r = \pi$ (180°), it then follows that

$$E_S = 2E_R \sin\left[\frac{2\pi h_R h_T}{\lambda d}\right] \qquad \ldots (2.17)$$

This result shows that at close distances the field strength will oscillate if h_R or h_T are large, i.e. high, apart from the reflection coefficient problem. Alternatively written as signal strength

$$P_R = 4P_{direct} \cdot \sin^2 [2\pi h_T h_R / \lambda d]$$

in relation to the direct signal level above, and with distances $d > h$ (low angle of incidence), which reduces to

$$P_R = 4P_{direct} \cdot [2\pi h_T h_R / \lambda d]^2 \qquad \ldots (2.18)$$

Introducing the modified P_R back into (2.11) gives

$$P_R/P_T = G_T G_R \cdot [h_T h_R / d^2]^2 \qquad \ldots (2.19)$$

Written in logarithmic form the propagation loss thus becomes

$$L_{db} = 40 \log d_m - 20 \log h_T h_R \qquad \ldots (2.20)$$

This equation is of fundamental importance to terrestrial mobile radio. Note especially that it is an inverse fourth power law and is independent of frequency. (It is also the basis of CCIR TV and radio coverage data.) The distance d is now in metres, not kilometres, as in the free space equation, and this reminds us that the equation really only applies to small flat earth cells. However, when the operating distance is very small, as it can be in cellular, the above equation, and hence the propagation law, will break down. A signal envelope of the form suggested in Figure 2.9 could be observed.

The fourth power law applies as the mobile moves away from the base site; this implies a 12 dB fall in field strength for every doubling of the distance. Equation (2.20) also indicates that only range d and antenna height affect the signal loss, so the calculation is elementary.

Example

Same distance d	$= 1000 \text{ m} = 1 \text{ km}$
$h_T \times h_R$	$= 10 \text{ m}^2 \text{ (low antennas)}$

Therefore $L = 100 \text{ dB}$

and a P_T of 4 milliwatts is now required for a 1 km radio cell.

Example

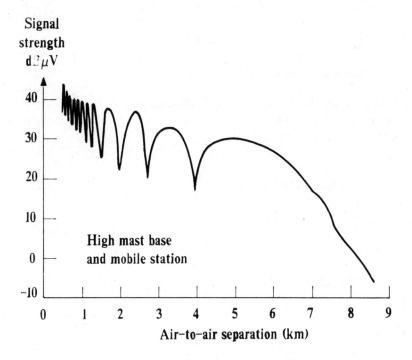

Figure 2.9   Received field strength variation observed because of high BS and close operational distance of MS

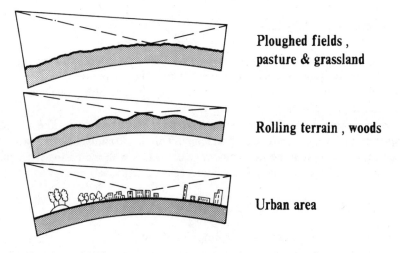

Figure 2.10   Types of 'rough ground' that can be encountered

Distance d   = 25 km

$h_T \times h_R$ = 100 m² (base station set high)

Therefore   L = 136 dB

the P_T required is now 16 watts.

Observations show that, apart from the ultimate effect of the earth's curvature over a long path (large radio cell), several terrestrial effects must be taken into account. These include

(i)   Surface roughness
(ii)   Line-of-sight obstacles
(iii) Buildings and trees, etc.
(iv)   Mountainous areas, etc.

The propagation loss equation is therefore amended to read

$$L = 40 \log d - 20 \log h_T h_R + \beta \qquad \qquad \ldots (2.21)$$

where β is the additional losses, lumped together; these are studied further below. If the *additional loss factor* β is a constant, this equation says that the radio cell (range equal to the radius) is circular. Because β, and h_R, can vary according to which direction one views from the transmitter, radio cells in general are not circular.

2.5.2   Rough ground model

Rough ground causes interference between the direct ray and reflected rays. Figure 2.10 shows the type of rough ground that may be encountered. Clearly the calculation of the received signal, even at one distance along one radial, is a complicated procedure.

The basic formula (2.21) is not complicated except that we do not know β. Also it appears to be independent of frequency. This is certainly not the case. All observations show that the radio coverage decreases with increasing frequency. Observations also show that the loss increases approximately as

f^n where n = 1 to 2 ;

this result can be introduced into an empirical formula.

Empirical model formulas are based on force fitting formulas to measured data. An example for VHF frequencies is

$$L_{dB} = 40 \log d_m - 20 \log h_T h_R + 20 + f/40 + 1.08\, L - 0.34H \qquad \dots (2.22)$$

where

f = frequency in MHz

L = land usage factor - the percentage of the test area covered by buildings of any type

H = terrain height differences between the T_X and R_X, (R_X terrain height − T_X terrain height)

Example

f = 160 MHz

L = 30%

H = 50 m (over a hill)

Therefore, excess loss, β = 20 + 4 + 6 + 15

= 45 dB

This additional 45 dB loss implies that a 125 W base station is now required to cover a path distance, or cell size, of 25 km.

An additional loss factor based on an urbanisation factor U can also be introduced. These models are derived with the aid of measurements taken. Actual received signal strength is compared to that predicted by the plane earth equation alone, and the difference (excess clutter factor) found. The final model is therefore composed of the plane earth equation, plus a clutter factor which is a best fit equation based on the factors considered most likely to increase propagation loss.

2.5.3 CCIR standard model

No single model can be expected to fit all conditions and all locations, but the CCIR has suggested a model, based on a long series of observations, known as the CCIR empirical formula for urban areas. It has the following form

$$L_{dB} = 69 + 26 \log f_{MHz} - 14 \log h_T$$
$$+ (45 - 6.5 \log h_T) \log d_{km} - A\,(h_R) \qquad \dots (2.23)$$

where

$$A(h_R) = (\log f - 0.7)\, h_R - (1.6 \log f - 0.8)$$

Note the fourth power d law holds exactly for $h_T = 5.6$ m.

Cellular Radio

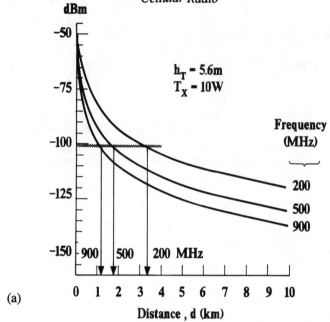

Figure 2.11   Signal strength versus distance according to (2.23). (a) BS height
5.6 m and P_T = 10 W (b) BS height 50 m and P_T = 100 W h_T = 1.5 m

This equation is plotted as received field strength in Figure 2.11 for various frequencies of operation and assuming the base station powers indicated. By drawing the -101 dBm (50μV/m) threshold line on the signal level scale, some idea of the urban radio range can be read off the distance scale.

Figure 2.11(a) is for $h_T = 5.6$ m (fourth power law). Figure 2.11(b) is for $h_T = 50$ m, a much higher placed BS. The signal level performance variation close to the BS is clearly not representative, as discussed, but the cell size variation with frequency is well illustrated.

The above empirical model does not consider penetration of radio waves into buildings. Penetration certainly takes place, the *building penetration loss* being a complex function of frequency and window sizes, etc. A 10 dB reduction of signal level can generally be expected.

2.6 Cell site coverage assessment

A reasonable estimate of the coverage area of a transmitter can be achieved by two means: (i) using a repeater technique (as used in conventional and private land mobile radio systems) will provide an estimate of the radio coverage (cell boundaries) and also does not require the transmitter to be continuously operating; (ii) using a field strength measuring receiver attached to a distance or location monitor. New instruments, handportable or vehicle mounted, can achieve very impressive field strength maps. For digital cellular systems a map showing the bit error rate topography of each cell site is probably more important. This requires sending packets of data over the radio path, say mobile-to-base, and comparing the received data sequence to the expected data in order to determine the BER at the particular location. Poor BER performances do not require long sequences in order to make a measurement, unlike a good BER performance which implies say, at least 100,000 bits of data.

2.7 Computer prediction techniques

For a detailed theoretical field strength prediction, radii from the transmitter (BS) site are laid out on a detailed topographic map of the area (in the UK available from Ordnance Survey). Height data to within 10 m at every 10 m or 50 m intervals, in the form of a grid, can now be purchased from the appropriate authority. Figure 2.12 outlines the concept. Choosing a radial at a specific angle, the ground contour can be averaged from the survey data set out in the form {x, y; h}.

If the ground above the radial under consideration is level, or perhaps just urbanized, the signal level path loss formula described above could be used. However, if conditions are hilly, a hill can present a considerable obstacle between a BS and a MS; such as suggested in Figure 2.13.

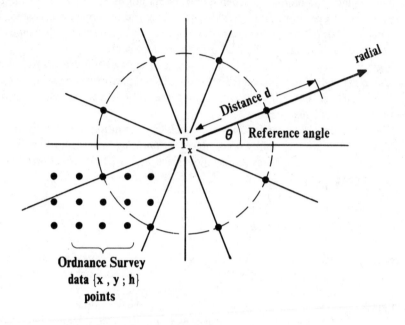

Figure 2.12   Radial lines for planning cell coverage overlaid on topographic
                data points

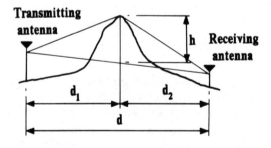

Figure 2.13   Obstacle between transmitter and receiver; the so-called knife
                edge model

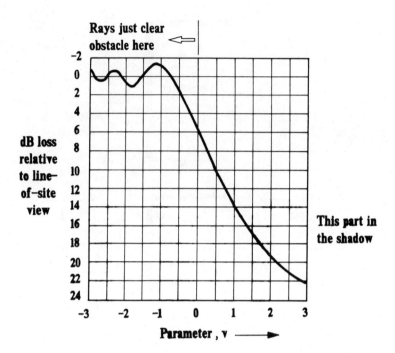

Figure 2.14   The additional loss due to an obstacle in the path of a radial

Radio waves do not form an absolute shadow at the other side of an obstacle; rather, zones of radio wave influence are formed in the shadow of the obstacle, known as *Fresnel zones*. Signals found in the shadow and their relative amplitude can be calculated, as shown in Figure 2.14. The loss is determined by a dimensionless geometric term v, given by

$$v = h\left[\frac{2}{\lambda}\left(\frac{1}{d_1} + \frac{1}{d_2}\right)\right]^{\frac{1}{2}} \qquad \dots(2.24)$$

Clearly UHF is more affected by such obstacles because λ is small, etc. The results can be fed automatically into computerised field strength programs when and where such an obstacle appears along a radial. A typical program structure is shown in Figure 2.15. The functions discussed above will be recognised, and supported by a composite diagram. In addition, two important additional features

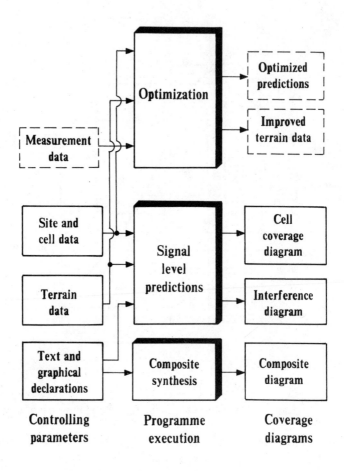

Figure 2.15   Management arrangement within a computerised cell site radio
               coverage programme

are the calculation of signals from possible surrounding cells, and also the optimization of the terrain data and model prediction by force fitting to measured data, as described above.

Finally, by way of illustration a composite cell site prediction result is shown in Figure 2.16. This comes from a Vodafone engineering plan for part of the coastline in the vicinity of the city of Blackpool in the north of England.

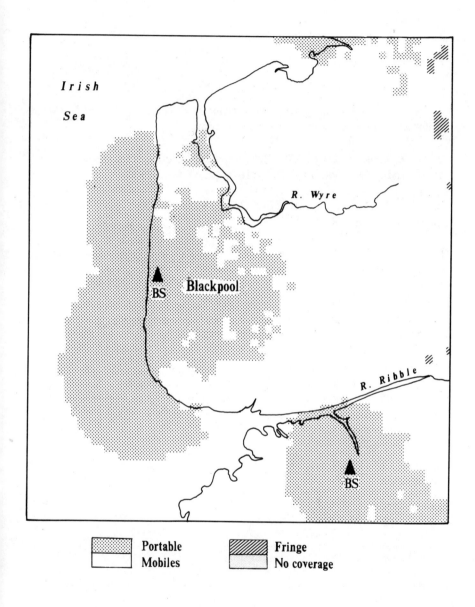

Figure 2.16   Computer predicted coverage for the city of Blackpool and the
surrounding coastline and inland region, assuming a BS in the
position shown. [Diagram supplied by Vodafone Engineering
Services, UK]

Further reading

CCIR (1982). *Recommendations and reports, XV Plenary Assembly,* Geneva, Volume V, (Propagation in non-ionized media)

Edwards, R. and Durking, J. (1969). 'Computer prediction of service areas for VHF mobile radio networks', *Proc IEE,* 116, Sept, pp 1493-1500

Griffiths, J. (1987). *Radio Wave Propagation and Antennas,* Prentice-Hall International Ltd, UK

Holbeche, R.J. (1985). *Land Mobile Radio Systems,* Peter Peregrinus, IEE Press, UK

IEEE Vech Tech report on propagation (1988). 'Coverage prediction for mobile radio systems operating in the 800/900 MHz frequency range, *IEEE Veh Tech,* VT-37, Feb, pp 3-70

Lee, W.C.Y. (1982). *Mobile Communications Engineering,* McGraw-Hill Book Co, USA

Matthews, P.A. (1965). *Radio Wave Propagation VHF and above,* Chapman and Hall, London, UK

Parsons, J.D. (1992). *The Mobile Radio Propagation Channel,* Pentech Press, London, UK

3 Cellular Radio Design Principles

3.1 Analog cellular frequency allocation plans

In its simplest concept radiotelephones are set up by assigning one pair of channels to each user or phone. As in broadcasting, a channel is defined by its centre frequency and bandwidth. This scheme is known as frequency division multiplexing (FDM) and works very well. It is the basis of all present day analog cellular schemes. This channel planning concept is shown in Figure 3.1.

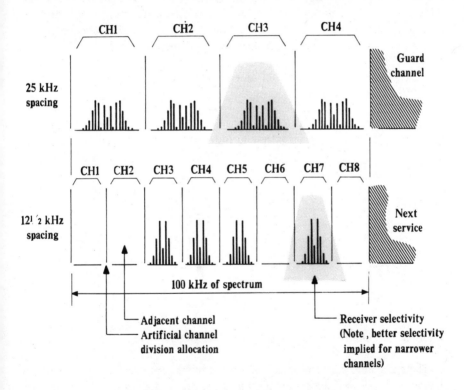

Figure 3.1   Channel arrangement within a 100 kHz band of spectrum: four 25 kHz channels, or eight 12.5 kHz channels are possible

This plan shows, either the spectrum at the base station transmitter, or the spectrum seen by the mobile station receiver. For duplex operation, the mobile also needs to transmit and the base station receive (on multiple receivers). In order to

59

avoid the receiver's seeing the same transmitted frequency, or the base receiver for a distant mobile being interfered by the transmission from a nearby mobile, the transmit and receiver frequencies must be well separated. The frequency difference between the channel pairs is usually 45 MHz, and is within the performance capability of the duplexer shown in the mobile circuit outline drawn earlier as Figure 1.3. The channel carrier spacing is normally 25 kHz, at least in the UK.

A simplified view of the band plan for the UK TACS system is shown in Figure 3.2. The radio spectrum assigned is basically 890-915 and 935-960 MHz. This therefore allows one thousand channels for the service. The forward paths are the higher frequencies (because as we noted in Chapter 2 higher frequencies have more propagation loss) and refer to the BS transmit frequencies. The return paths are the lower frequencies and refer to the MS transmitter frequencies. Since the edge of this band is 890 MHz, the actual first transmit frequency specified will be 890.0125 MHz, i.e. 12.5 kHz up from 890 MHz, as explained by Figure 3.1. Likewise, considering channel 27 for example, which is allocated as a control channel for the base station, the signal here will appear on the frequency

$$F_{27} = 935 + 26 \times 0.025 + 0.0125$$
$$= 935.6625 \text{ MHz}$$

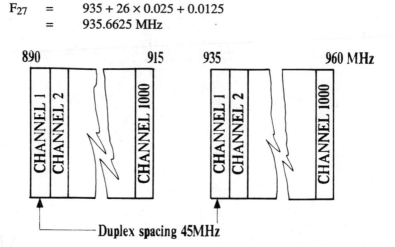

Figure 3.2   Outline allocation for TACS frequencies and channels

The channel assignment is actually made more complicated as a *duopoly* of operators is allowed access to the radio spectrum channels. Calling the two operators A and B respectively, the general arrangement for sharing can be illustrated by reference to the USA AMPS assignments, Figure 3.3. The frequency limits are not the same as indicated in Figure 3.2 (the regional variation principle described in Chapter 1), nor is the channel bandwidth the same; 30 kHz is employed by AMPS. The result is that one has (the original) 666 channels, plus 166 additional channels from a more recently assigned 5 MHz of spectrum. Note that the channel numbering scheme goes up to CH799, then stops, and begins again at the lower frequencies for CH991 to CH1023. The control channels, twenty-one per operator, go from channels 313–333 and 334–354, respectively.

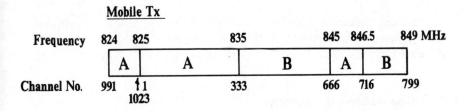

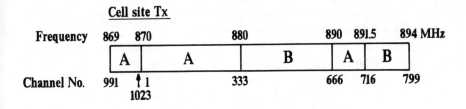

Figure 3.3   Current AMPS channel allocation arrangement

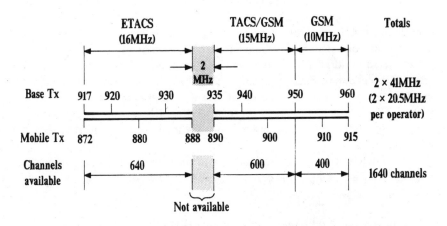

Figure 3.4   Current spectrum allocation to cellular radio in the UK

The allocation of spectrum to cellular radio in the UK, at present, is shown in Figure 3.4. Notable differences from Figure 3.2 are worth highlighting. Firstly, extended TACS, or ETACS, provides a further 320 channels per operator (Vodafone and Cellnet), by going to the lower frequency limits of 917 and 872 MHz, respectively. On the other hand, the top 10 MHz of each band is now assigned to the scheduled pan-European digital cellular network (GSM), so the upper bounds are now 950 and 905 MHz respectively. Table A.1 in Appendix I contains this information.

Again, it needs to be recalled that a number of channels need to be dedicated as control channels. Like AMPS, twenty-one are assigned per operator; for TACS, these are channels 23 to 43 and 323 to 343 respectively.

3.2 Base station site engineering

It will be clear that the mobile unit will have to be very frequency agile and also carry an antenna which can both receive and transmit over a fairly wide band of frequencies; that is, the antenna must be broadband especially with regard to its voltage standing wave ratio (VSWR) on transmit.

Any base station at a particular cell site will have a very different task, however; not only will it be receiving many signals, channel by channel, it will also need the capability of transmitting simultaneously on many separate channels, though not quite as adjacent as in Figure 3.1. As we shall note later, digital cellular radio minimises this particular problem to some extent (at the expense of another), but this multiplicity of operation is a very real requirement.

Although separate common transmit and receive antennas could be used at the base station, with gain especially on the return path, because the two antennas would be in close proximity, they could to all intents and purposes be regarded as one, due to the close coupling involved. To separate the signals a methodology known as antenna site engineering is used, illustrated by Figure 3.5.

Spectrum dividing filters, as shown, differentiate between the forward and return channel bands. Each transmitter would have an associated multicoupler in order that there was no return transmitted power, between the channelized transmitters. The BS receivers would have to have a very linear common pre-amplifier, to ensure low intermodulation between very unequally sized received carriers, and a low noise figure.

The channels in use at any particular base station will not be adjacent channels, but separated according to the cell cluster pattern arrangements described below.

3.3 The concept and benefits of channel sharing

The combination of the mobile station being frequency agile and the allocation of many channels in a band means that each mobile has *access* to many channels. In analog cellular this is on a channel-by-channel basis, termed FDMA; in digital cellular described later, a band-by-band allocation is used and is termed narrow band TDMA.

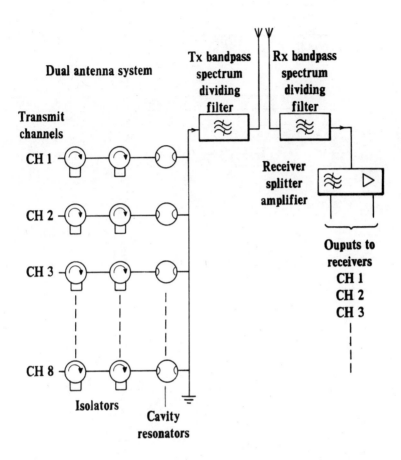

Figure 3.5　　Outline of base station antenna site components

Figure 3.6 illustrates the arrangement which applies to either the forward or the return path between the BS and MS. Two observations may be made; firstly, more subscribers than channels appears to be possible, especially if each call (message) is short and the subscribers make use of the network randomly. This is known as the principle of *trunking*; demonstrated by the drawing in Figure 3.7. The principle of trunking comes from telephony and is a scheme which allows the subscribers to have access to all available channels and hence have a many-fold increase in the likelihood of a successful connection. A figure for the trunking gain can be calculated when facts about the subscriber traffic behaviour are known.

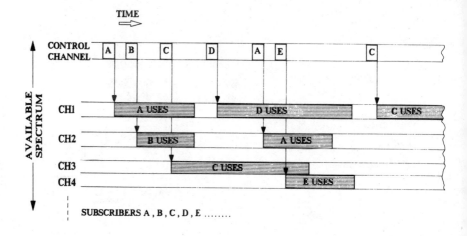

Figure 3.6   Demand allocation of voice channels

CONCEPT OF TRUNKING GAIN

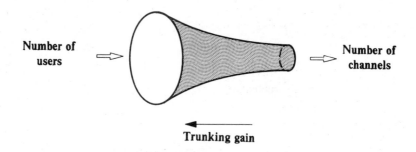

Figure 3.7   A conceptual view of trunking gain

In a trunked scheme a larger number of mobiles can be given a more acceptable grade of service than could be supported on an equivalent number of single channel systems. Conversely, the same number of mobiles on single channel systems could be given a better grade of service on a trunked system; or, the same number of mobiles could be given the same grade of service on fewer channels.

This thus saves the number of channels needed. The reliability of the system is also very much greater. The loss of one base station channel merely degrades the service; to a single channel user this would mean total loss of communication.

We need to define *grade of service* (GOS). This is a measure of the probability that a successful connection (a call) will not be achieved (a somewhat back to front definition). Telephone companies look for a GOS in the range of value 0.01; implying that only one call in 100 is lost. This is a theory very much at the heart of telephone traffic performance and is described in more detail in Chapter 9. For the present we just need to accept the obvious advantage of many channels sharing, or trunking, as depicted in Figures 3.6 and 3.7. To put numbers on the gain requires much theory and assumptions.

The second point from Figures 3.6 is that the operation of a control channel is needed to manage the channel (frequency) allocation as requested by each MS. This leads to the complicated issue of signalling in the cellular network (next chapter), but also one notes that the control channel could itself become blocked, due to over use, and hence limit the number of subscribers entering a radio cell even if it were allocated a large quantity of voice channels.

Clearly, channels must be allocated with a minimum of delay. In some specifications, 1200 bps fast frequency shift keying has been chosen as a compromise between speed and reliability within a 12.5 kHz channel separation. The TACS cellular system employs 8 kbps direct modulation. (A channel separation of 50 kHz is now needed to accommodate this.)

The protocol used must ensure full reliability of correct connection in the mobile environment. The signalling telegrams themselves have powerful error-detection capability so that significant falsings will not occur. However, many messages are likely to be rejected in bad signal conditions and therefore the protocol must be designed around acknowledgements and retries. A balance between no retries and an excessive number must be agreed which does not overload the control channel, but in cellular the traffic channels are usually limiting because no limit on conversation length is applied.

The inclusion of queuing in a system design is of interest. If a system has become very busy the only means of access to a traffic channel, if there is no queuing, is by continuous retries. This may be accomplished manually or automatically or by a combination of both. These retries will tend to block the control channel and at the moment a traffic channel becomes available contention could be so great as to require time to reallocate it. This reduces the efficiency of the system and could generate frustration to the user. Secondly, the user may have a more favourable perception of the system when he obtains a queued message. Also, emergency and priority calls can be more readily processed.

Notice that a certain amount of privacy is built into the frequency agile FDMA technique, because it is not known on which channel a particular subscriber may appear. Also the system can be expanded by adding more channels at a particular cell site, if necessary.

To recapitulate, in any cellular system the operation has recourse to a great many voice channels at a particular location, the more so, the more effective channels appear to be on offer because of the trunking gain. However, once the channels (or frequencies) have been allocated, they are not available; the subscriber has to move to another site, or wait. This brings us to the principle of cell planning which we shall now discuss.

3.4   Multiple cell plan

The basic shortcoming of the single large radio site is the imbalance of power levels at the centre and the edge of the cell. For example, suppose one could cover a 1 mile radius cell with a 100 mW transmitter power. To cover a 10 mile radius cell would require 100 W transmitter power - because of the inverse fourth power propagation law alone.

The only way to achieve coverage with lower powers is to arrange coverage with a subset of small cells, as in Figure 3.8.

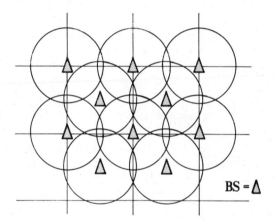

Radio cell plan set on national grid

Figure 3.8   Radio cells laid out on a national grid network. (This arrangement is to be used for the European terrestial flight (cellular) telephone system)

The cellular concept was described in some detail in Chapter 1; Figure 3.8 shows a more practical layout arrangement, at least over a geographically flat area. There is a good deal of overlap of coverage, but this does not matter too much since a different set of frequencies is used in each cell. However, it is clear from the overlaid drawing of a regular hexagon and a circle in Figure 3.9 that the hexagonal shape is a much more attractive geometrical shape to use for planning multiple cells than a circle. Other possible geometric shapes are triangles (used by broadcasters) and squares. in each of these cases the shapes can be 'tessellated', that is, fitted together exactly.

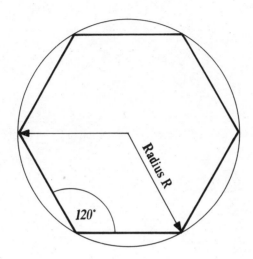

Figure 3.9   The regular hexagon coverage area as compared to a circle. The cell radius R is indicated

3.4.1 Cell structure geometry

Hexagons may be packed side-by-side (no overlap necessary) and examination of a set of hexagons shows that they can be packed in *clusters* such that no two similar cells are adjacent, i.e. use the same set of frequencies. The cluster size is designated by the letter N. N is determined by the condition:

$$N = i^2 + ij + j^2$$

where i, j = 0, 1, 2, etc.

Thus only the cluster sizes 3, 4, 7, 9, 12, etc, are possible.

These repeat patterns are illustrated in Figure 3.10. An individual cell may be sectored, in particular by 120°, as evident from the geometry of Figure 3.9. This means that a 9 cell cluster can also be regarded as 3 clusters of 9 sectored cells, called a 3/9 cell cluster. Other examples are shown in Figure 3.10.

Normal   3/9 cell cluster
3 site cluster   (sectored 3 cell cluster)

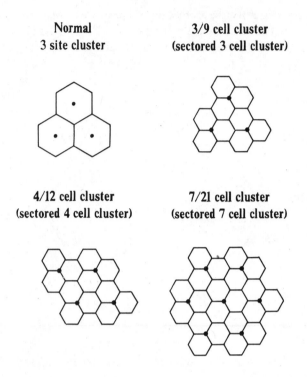

4/12 cell cluster   7/21 cell cluster
(sectored 4 cell cluster)   (sectored 7 cell cluster)

Figure 3.10   Cellular repeat patterns. The base station locations are indicated by
               the black dots

3.4.2   Reuse distance

The cell cluster size has two significant attributes. Thus if one focuses on the popular 7-cell cluster arrangement, redrawn in Figure 3.11, it is firstly noticed that the allocation of frequencies into seven sets is required. Thus if 210 frequencies were available, this would mean that only 30 channels per cell could be assigned, one or more of which would be needed as a control channel, so that at least 14 control channels would be needed for the cluster. For duplex operation this pattern would be repeated in the two allocated bands.

The second important aspect of Figure 3.11 is the mean distance between cells using the same frequency set. This is called the *mean reuse distance* D. Because of the geometry of hexagonal cells, D is related to the cell radius R, and the ratio of D to R, called the *reuse ratio,* is a function of cluster size, and

$$\frac{D}{R} = \sqrt{3N}$$

$$\ldots (3.1)$$

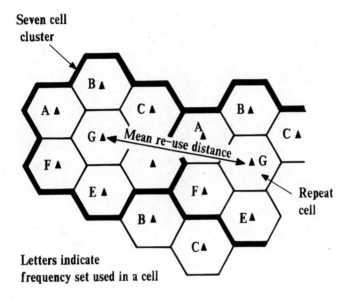

Seven cell
cluster

Letters indicate
frequency set used in a cell

Figure 3.11   Seven-cell repeat pattern showing the mean reuse distance
                   between cells

Thus for a 7-cell cluster of 2-mile radius cells, the repeat cell centre would be
9.2 miles away.

The distance away of a similar on-frequency transmitter will cause co-channel
interference to a mobile in its rightful cell. For a 7-cell cluster there could be up to
six immediate interferers, as shown in Figure 3.12.

Assuming the fourth power propagation law, an approximate value of the
carrier C to interference I ratio is

$$\frac{C}{I} = \frac{C}{\Sigma 6 I\text{'s}} = \frac{R^{-4}}{6D^{-4}}$$

assuming that the interferers contribute equally.

Using $D/R = (3N)^{\frac{1}{2}}$

$$\therefore \quad \frac{C}{I} = \frac{1}{6}(3N)^2 = 1.5N^2 \qquad\qquad \dots (3.2)$$

i.e. the C/I ratio is a function of the cluster size; it is designated C_i. (Note that the
result is an approximation for values other than N = 6, but is adequate for the
discussions which follow.)

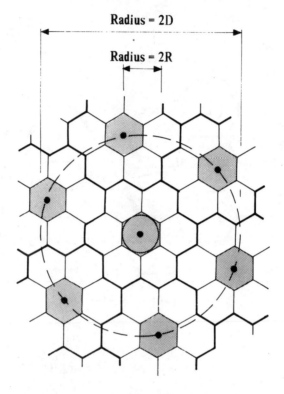

Figure 3.12   The geometry associated with interfering cells using a 7–cell
cluster pattern

Example   Suppose N = 7, then C_i ≤ 73 or 18 dB.

This appears very adequate, but two other facts have to be taken into account:

(i)   *Adjacent channel interference* from channels in adjacent neighbouring cells.
      This is worse in small cell clusters.

(ii) *Multipath fading* may weaken C as against I, discussed later.

However, it is useful at this stage to draw up a table comparing the properties
of various cluster sizes. This is shown as Table 3.1.
    It is also of interest to plot the number of channels per cluster and the C_i ratio versus
cluster size N, and note the step function nature of the result, shown in Figure 3.13.

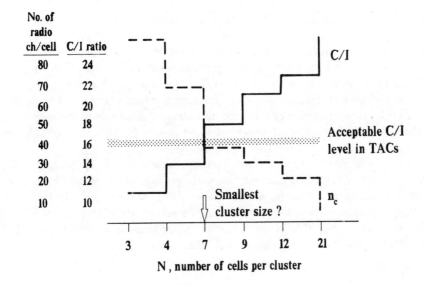

No. of radio ch/cell	C/I ratio
80	24
70	22
60	20
50	18
40	16
30	14
20	12
10	10

N, number of cells per cluster

Figure 3.13   Number of radio channels per cell (out of 300) and C/I ratio (dB) plotted against cell cluster size N

Table 3.1   Influence of cluster size on individual cell parameters

Cluster size N	(a) reuse ratio	(b) max no. of CH per cell	(c) co-channel C_i (dB)	(d) no. of subs. per cell
3	3	93	11	2583
4	3.5	69	14	1840
7	4.6	39	18	937
9	5.2	31	21	707
12	6	23	23	483
21	7.9	14	28	245

(a)   using (3.1)
(b)   300 channels per operator assumed available, of which 21 must be control channels
(c)   using (3.2)
(d)   for calculation here, see Chapter 9

The parameters shown in the Table 3.1 are independent of the cell size R, which is assumed to have the *same value* for all cells in the cell plan. The actual reuse distance will depend on the value of R chosen, i.e. for 2 km cells the second column

is half the reuse distance in km. The third column depends on the frequency
allocation plan, and applies more or less to AMPS and TACS. The C_i ratio is in
decibels in the fourth column and is independent of cell size. Large cells fare no
better than small cells, which at first sight does seem strange, but of course large
cells need more effective and positioned radiated power. The final column requires
calculation of the trunking gain, which is postponed until Chapter 9. The figures
assume a grade of service of 0.01 and an average calling rate of the individual
subscribers. The apparent advantage of small cluster sizes may not actually be
necessary except in the centre of large busy cities.

In the UK TACS system N was initially chosen as 7 or 12 because of C_i
considerations. With 300 channels available per operation, of which 21 were
dedicated as control channels, this leaves the 39 or 23 channels per cell. The voice
channels in cells are chosen to be as far away from those in their adjacent cells, from
the channel set available in the frequency plan, because of adjacent channel
interference and bandwidth overlap problems.

3.4.3   Cell splitting

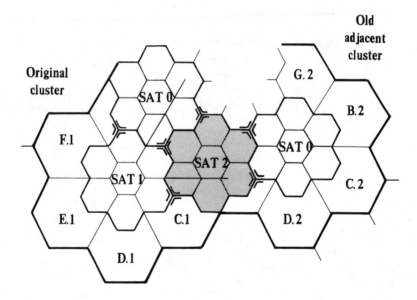

Figure 3.14   Cell splitting procedure

For a given value of N, the capacity of a system may be increased by reducing the
size of the cells so that the total number of *channels available per unit area* is
increased. In practice this is achieved by the process of 'cell splitting', where new

base stations are established at specific points in the cellular pattern, reducing the cell size by a factor of 3 or 4. By repeatedly splitting cells, the system capacity can be tailored to meet the traffic capacity requirements demanded by customers, in areas, from low traffic rural areas, say, where the cells may be 20 km radius or more, to high traffic central urban areas, say, where cells may be as small as 1 km radius. In practice, the variation in propagation, particularly in urban centres, and the accuracy in position to which base stations can be located, are the main factors limiting the minimum cell size.

An example of cell splitting is shown in Figure 3.14. When this exercise is carried out the supervisory audio tone (SAT) coding of the newly created clusters must also be revised. This was discussed to some extent in Chapter 1, in particular Figure 1.15. There are three SAT tones; keeping the centre of the original cluster with its original SAT tone, Figure 3.14 indicates which tones are applied to the new smaller clusters. Similar considerations will also apply to the digital colour codes (DCC) of which there are four.

3.4.4   Sectorization

Although in theory, as the cells are split into smaller sizes, the interference received from reuse cells, indicated in Figure 3.12 and Table 3.1, in irregular propagation in dense urban areas, and non-circular cell shapes, leads to increasing interference being received from the surrounding cells all using the same channel set. One way of reducing the level of interference is to use directional antennas at base stations, with each antenna illuminating a sector of the cell, and with a separate channel set allocated to each sector. There are two commonly used methods of sectorization, using three 120° sectors or six 60° sectors, both of which reduce the number of prime interference sources. The three sector case is generally used with a seven-cell repeat pattern, giving an overall requirement for 21 channel sets, as shown in Figure 3.10. The improved co-channel rejection in the six sector case, however, particularly the rejection of secondary interferers, results in a four-cell repeat pattern being possible, but needs an overall requirement of 24 channel sets.

A disadvantage of sectorization is that the larger number of channel sets required results in fewer channels per sector, and thus a reduction in trunking efficiency. This means that the total traffic which can be carried for a given grade of service is reduced. However, the capability to use much smaller cells through sectorization outweighs such drawbacks, and the end result can be a higher capacity system.

3.4.5   Other cell patterns

The overlaid cells concept is shown in Figure 3.15. It allows further reuse of frequencies at each site, provided that they are only used by mobiles within a

Cellular Radio

smaller radius than that of the macrocell, thus providing additional capacity at the centre of a cell. It is important that the cell site is located where the peak traffic occurs. The overlaid cells use frequency groups that are already allocated to normal adjacent macrocells. The reuse distance for the overlaid cell appropriate to a seven-cell cluster can thus be maintained and hence the quality of service is approximately the same as that of the main cell plan.

For illustration, the London cell plan as used by the UK Vodafone network is shown in Figure 3.16. This illustrates the use of the techniques just described to achieve adequate traffic capacity. (The diagram should be compared to Figure 1.7.) It demonstrates how cell layout strategy has developed in order to meet increasing numbers of subscribers on the network ; in London it is now approaching half a million.

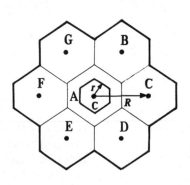

Figure 3.15   An overlaid cell
                arrangement

Figure 3.17   The Stockholm ring plan
                and the first expansion
                pattern

Both TACS and AMPS are not particularly good in respect of adjacent channel interference because the signalling tones (see next chapter) are fast frequency shift keying waveforms with a high deviation and, as discussed in Chapter 6, the modulated waveform adjacent channel spectrum is not well controlled. This is not the case with the lower deviation, slower speed signalling systems such as NMT 450 and NMT 900 (see Table A.1). A cell layout strategy known as the Stockholm ring can be employed. The central site can now have access to all the available channels which are sectored in 60° co-sited sectors. Expansion away from the centre is demonstrated by Figure 3.17. This strategy could well be suitable for the walled European Roman type of city, such as Avignon in France, Chester in England and Solothurn in Switzerland. If a large number of channels are available to the cellular service being contemplated, a hexagonal layout pattern is most likely to be employed however, as expansion into the surrounding rural areas becomes simpler. The cell pattern can also be effected by changing the handover parameters,

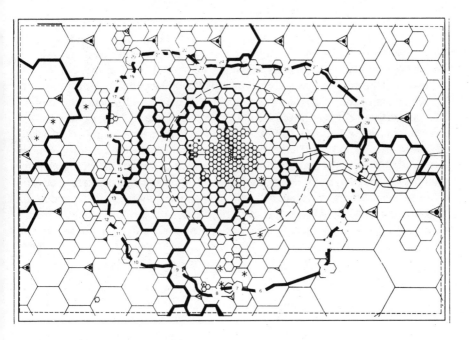

3.16 Current Vodafone London cell plan arrangement

discussed below. As perhaps could be ascertained from Figure 3.8, this is a strategy of assigning the region of overlap between cells to one or the other cell, hence reshaping the layout pattern. This option is under the command of the mobile switching centre (MSC) that oversees a particular cluster of cells.

3.5 The cellular system

The basic concepts of frequency planning and reuse, and the control of co-channel interference, are equally applicable to private mobile radio systems and indeed also apply to TV and radio broadcasting. What is different with cellular is that the individual base stations are interconnected to form a complete system, offering continuous coverage with a minimum user inconvenience. There are two key features of cellular systems which make this possible, mobile location and in-call handover.

3.5.1 Mobile location

When an incoming call is received for a mobile station, the call has to be routed to the cell where the mobile is located so that the call can be connected. One way of finding the mobile would be to transmit a calling message (page) for the mobile on every cell site in the network. However, with several hundred cells and hundreds of thousands of mobiles, the signalling capacity required would clearly be too high. Instead the cellular network is split up into a number of location areas, each with its own area identity number. This number is then transmitted regularly from all base stations in the area as part of the system's control information. A mobile station, when not engaged in a call, will lock on to the control channel of the nearest base station and, as it moves about the network, will from time to time select a new base station to lock on to. The mobile station will check the area identity number transmitted by the base station, and when it detects a change, indicating that the mobile has moved to a new location area, it will automatically inform the network of its new location by means of a signalling interchange with the base station. In this way the network can keep a record (registration) of the current location area of each mobile, and therefore be able to call the mobile only within that area.

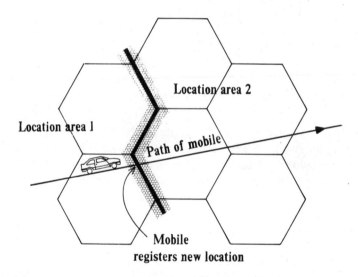

Figure 3.18   Mobile location registration

3.5.2 In call handover

When a mobile station is engaged in a call, it will frequently move out of the coverage area of the base station with which it is in communication, and unless the call is passed on to another cell, it will be lost. The system continuously monitors the signals received from mobiles engaged in calls, checking on signal strength (and quality). When the signal falls below a preset threshold the system will check whether any other base station can receive the mobile at better strength and, if this is the case, the system will allocate a channel for the call at the new base station, and the mobile will be commanded by a signalling message to switch to the new frequency. The whole process of measurement, channel allocation and handover may take a few seconds to complete, but the user will only notice a break in conversation of 200-300 ms as the handover itself is carried out.

Effective and reliable handover is not only highly desirable from the user's point of view, but essential in the control of co-channel interference and thereby the maintenance of the cell plan, particularly as the cell size is reduced. A mobile operating in a non-optimum cell will, in effect, be operating outside the cell designated for that area. In other words, the cell boundary will have been altered beyond its planned limit, and this will give rise to levels of co-channel interference above that planned for the adjacent system.

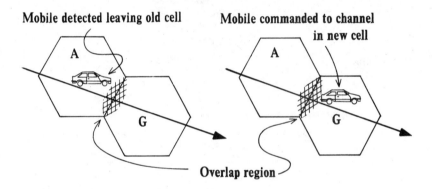

Figure 3.19   In call handover operation

A further means to control co-channel interference is that of mobile power control. So long as the base station is receiving a signal of adequate strength from a mobile, there is no need for the mobile to be transmitting extra power, so the base station can command the mobile to reduce power by sending a signalling message. Clearly, by reducing a mobile's power, the likelihood of its causing interference is also reduced, thus helping to control interference levels.

The radiated power of the base station is kept constant, however, since this defines the cell size. The cell edge signal strength for AMPS/TACS follows recommendations based on experience. These are:

Urban cell	60 dBµV/m ~ 80µV at Ae
Suburban cell	39 dBµV/m ~ 7µV at Ae
Rural cell	34 dBµV/m ~ 4µV at Ae

3.6 The cellular network

In essence, all cellular networks have a similar structure, being complete telephone networks in their own right, with dedicated exchanges within an interconnected network, and with base stations connected to the exchanges. There are, however, many ways of planning a cellular network in practice, the optimum arrangement for any particular application being dependent upon the capacity required, cost of implementation, capabilities of the chosen manufacturer's equipment, etc. As an illustration of the network design, Figure 3.20 shows sets of seven-cluster cells connected in turn to their command centre, that is MSC. Each mobile switching centre will have a home location register (HLR) and visitor location register (VLR) associated with it, as described in the introductory chapter.

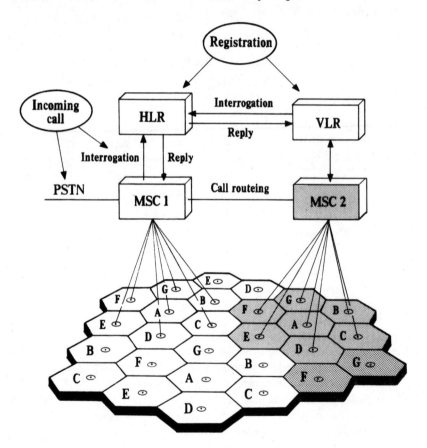

Figure 3.20   A conceptual diagram of the fixed network supporting the cellular radio layout

3.6.1 Base stations

For much of the network, base stations are organised in a 7-cell or 12-cell repeat pattern with omnidirectorial coverage from each base station. Most base stations have between 20 and 30 voice channels, with one signalling channel carrying all paging and access functions are per cell. The signalling or control channel transmitters and receivers are fully redundant, both operating in a work/standby mode. Voice channels are non-redundant, and any faulty channels can be taken out of service, only slightly worsening the grade of service of the cell.

The voice channel and signalling channel transmitter outputs are combined using high-Q cavity resonators and fed to co-linear antennas of 9 dB gain. The maximum effective radiated power (ERP) of each channel is 100 W, but in many cases a lower power level is used, as dictated by the co-channel interference requirements of the overall radio plan.

In the receive direction, many base stations are fitted with six directional antennas consisting of a colinear type mounted in front of a vertical corner reflector. These antennas have 17 dB gain, and are mounted at 60° intervals around the compass, starting at due north. A preselector/preamplifier is connected to each antenna and the six outputs are connected to a switching matrix which allows any one of the voice channel receivers to use any pair of antennas at any time. A pair of antennas is connected to the voice channel receiver, so that diversity can be used to minimise the effects of fading on the received signal (see Chapter 5).

In order that the best antenna is always connected to each voice channel, the base station has a scanning receiver which monitors the signal level on every channel used by the base station, via every antenna, every few seconds. The results of these measurements are then used by the base station controller to operate the antenna switching matrix. The scanning receiver also carries out measurements of signal strength for both hand-in and hand-out from the cell and for mobile power control. Like the signalling transceiver, both scanning receiver and base station controller are fully redundant.

Using directional antennas for receiving, even when the base station has a nominal omnidirectorial coverage, brings a number of advantages. The high antenna gain and the use of diversity improves receive performance, particularly for handportables, compensating for the power difference between base station and mobile. Co-channel interference in the mobile-to-base station direction is reduced since the base station is only looking one way, and therefore seeing fewer interferers. Also, handover processing is improved since the system can establish the direction of the mobile from the current base station, and therefore indicate more closely the next best cell for handover.

In some urban areas, particularly large cities, where customer numbers and usage demand a very high capacity service, base stations can be arranged in a four-cell repeat pattern with six sectors per cell. Each of the six sectors in a four-cell cluster is allocated its own set of voice channels. Control channels, however, are allocated on a one-per-base-station basis, with one signalling channel transceiver in the base station operating as a resource shared by the six sectors. When a mobile initiates a call, the base station controller establishes in which sector's coverage area the mobile is located, and allocates a voice channel in that sector. During a call, if the mobile moves to a different sector, the base station controller can within a few seconds carry out a sector-to-sector handover to ensure that the mobile is being covered by the best sector.

3.6.2   Mobile switching centres

A national switching network can consist of over 20 mobile switching centres, but in fewer locations. The MSCs are digital exchanges with a distributed control architecture, especially adapted for operation in the cellular environment.

Base stations are connected to the switching centres by digital (2 Mbps) leased lines. The switching centres are also linked together with digital circuits forming a fully interconnected network. The signalling between base stations and switches, and between switches, is usually proprietary in nature, and is carried in time slots on the digital circuits.

The MSCs are connect to the PSTN at a large number of locations in order to distribute the traffic load and to minimise the impact of any failures on call handling. Digital interconnection to the PSTN using the CCITT signalling system No. 7 is now used exclusively, having completely replaced the earlier digital/ analog interconnect with loop-disconnect signalling, as existed in earlier fixed telephone networks.

3.6.3   Other services

The network, as with other systems, offers fully automatic calling to and from telephones in the fixed network throughout the world, and also provides access to many of the services available on the fixed network, such as information services. In addition, the mobile switching centres support a range of 'vertical' services to complement the basic cellular service, such as:

- Call divert - all incoming calls are diverted to the specified number.

- Busy divert - incoming calls to a busy mobile are diverted to the specified number.

- No answer divert - incoming calls to a non-active (i.e. switched off) mobile, or to an unanswered mobile, are diverted to the specified number.

- Three-party conference calling - a third person can be brought into an existing conversation.

- Call waiting - an incoming call to a busy mobile is indicated by a tone to the subscriber, who can then pick up the second call, placing the first call on hold.

- Call barring - selective call barring can be invoked to prevent, for example, unauthorised international calling.

Other services provided by the network include a messaging service, a voice messaging service which is fully integrated with the cellular network and mobile numbering scheme, and private wire, which allows customers to take advantage of lower call charges by linking their private network directly to a mobile switching centre.

Further reading

Beddoes, E.W. and Germer, R.I. (1987). 'Traffic growth in a cellular telephone network', *Journ. I.E.R.E.*, 57, pp 22-26

Beddoes, E.W. (1991). 'UK cellular radio developments', *Elec. & Comms Eng. J.*, Aug, pp 149-158

Boucher, J.R. (1990). *Cellular Radio Handbook,* Quantum Publishing Inc, USA

Cellular mobile telephone system CMS 88', System description available from Ericsson Radio Systems AB, Sweden

Hughes, C.J. and Appleby, M.S. (1985). 'Definition of a cellular mobile radio system', *IEE Proceedings*, 132 Part F, Aug, pp 416 - 424

Jakes, W.C. (Ed) (1974). *Microwave Mobile Communications*, Wiley Interscience, USA

Lee, W.C.Y. (1989). *Mobile Cellular Communications Systems*, McGraw-Hill, USA

Thrower, K.R. (1987). 'Mobile radio possibilities', *Journ. I.E.R.E.*, 57, pp 1-11

4 Analog Cellular Radio Signalling

Unlike the fixed public telephone network (PSTN), the cellular radio telephone system has customers who 'roam' over the network. This is not the same concept as with the cordless telephone, for example, where the customer only moves about the same fixed base or for that matter, private mobile radio (PMR). The 'roaming' attribute means that the customer could be found anywhere within the network, which in the case of several systems can extend over national borders.

To provide this facility a very large amount of signalling overhead is required, some of which has already been described in the introductory chapter. This chapter describes the signalling procedures in much greater detail, and the procedures are also applicable to second generation systems.

4.1 Channel trunking needs

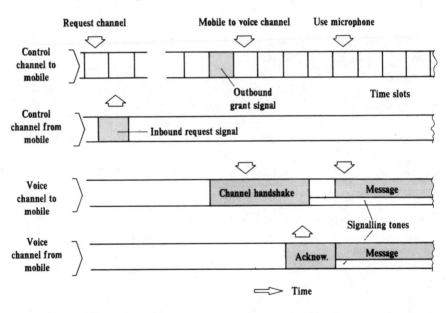

Figure 4.1   Protocol for granting a particular pair of voice channels to a mobile on request

Analog cellular sets up each mobile on a free channel in a cell when the *mobile station* (MS) calls, or is called by the local *base station* (BS). Figure 4.1 shows the strategy for the *control channel* (CC) to allocate a *voice channel* (VC) on request.

Continuous carrier transmissions ensue during the telephone call; in a busy cell all channels of the many transmitters and receivers at the base site could be in operation.

An example protocol between the BS and a single MS is shown in Figure 4.1. The four radio channels are those shown in Figure 1.14, namely the FCC, RCC, FVC and RVC respectively. The *forward* channel pair appear in the BS to MS frequency band; the *reverse* channel pair appear in the MS to BS frequency band, indicated above in Figure 3.4.

The signalling takes place during the request, handshake and connect periods and clearly constitutes much of the telephone call activity. In cellular, to this signalling must be added the identification, location and handover activities.

In TACS, channels 23-43 and 323-343 are the *dedicated control channels*, either the FCCs or RCCs, out of the one (two) thousand channels in the TACS/GSM spectrum allocation, namely Figure 4.2. These reserved channels are sometimes set aside as optional access channels and optional paging channels.

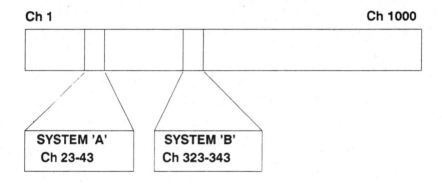

Figure 4.2   The dedicated control channels set aside in the TACS spectrum specification

The data on the *forward voice channel* (FVC) and *reverse voice channel* (RVC) is used for managing the calls. Data is transmitted on these channels before, after and during the call. The speech path is muted during the bursts of data, to prevent what would appear as interference to the speech circuit. Speech, data and supervisory tones are transmitted in a frequency shift keying format, each with particular modulation characteristics.

Speech is nominally modulated with a deviation of 5.7 kHz up to a maximum deviation of 9.5 kHz. This is a wide deviation compared to the channel spacing of 25 kHz and, to ensure that adjacent channels interference is kept to a minimum, adjacent channels are not used in adjoining cells, such as in Figure 3.11.

In comparison, the *supervisory audio tone* (SAT) is deviated by only a small amount, 1.7 kHz, but covers a cluster of cells, such as shown in Figures 1.7 and 1.15.

All other data is sent at 8 kbps and is modulated onto the carrier using FSK with a deviation of 6.4 kHz. The 8 kHz *signalling tone* (ST) is also modulated with 6.4 kHz deviation.

Before FSK transmission, the data is Manchester encoded, to ensure good synchronisation; that is the differential format shown in Figure 4.3.

In summary, the following signals actually occur in the message exchanging depicted in Figure 4.1.

• Voice modulation: FM, 9.5 kHz peak deviation, 5.7 kHz mean

• Signalling: 8 kbps Manchester encoding, direct FSK, or FM for ST, both 6.4 kHz deviation

• Supervisory tones: FM, 1.7 kHz deviation

• Signalling protection: 5/11 redundancy plus BCH block code

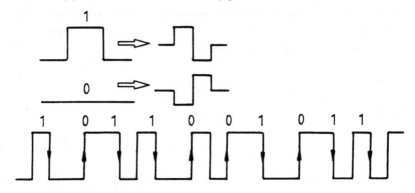

Figure 4.3   The Manchester encoding format for binary data

4.2   Equipment identity numbers

On delivery, every mobile is programmed with three numbers, namely,

ESN   Electronic Serial No. (32 bits)

MIN   Mobile Identity No. (34 bits)

AI   Area Identification (15 bits)

When the mobile is switched on, it reads this data contained in its memory, in particular, its MIN code.

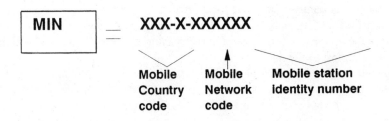

The MIN is a 34-bit binary number derived from a decimal number containing the mobile country code (MCC), the mobile network code (MNC) and a mobile station identity number (MSIN), which were the numbers discussed in section 11 of Chapter 1.

The *electronic serial number (ESN)* is of importance to the network operator, because it is the number which identifies the phone to the network, allows calls to be accepted or received, and also arranges for billing of the call charges to the operator (or owner) of the particular phone. Not long ago, the cloning of stolen equipment, in order to have the same ESN as an authentic paying customer took place, but procedures have now been introduced to curb this illegal activity.

The MIN number, as already discussed, refers the mobile to the supporting fixed network of the cellular radio system in which the MS is valid to operate. The important components of the fixed network in this respect are the *mobile switching centres* (MSC) and the locations registers, namely a *home location register* (HLR) and *visitors location register* (VLR).

Thus, when a mobile phone is switched on its data will be retrieved from a home location register (HLR) from somewhere in the network and stored in a visitors location register (VLR) on the switch serving the cells in the area where the phone is located. The HLR will note the identity of the current VLR and the fact that the mobile is active. Incoming calls for the mobile will interrogate the HLR, based on knowledge of the mobile's number and where each number is stored. If the mobile is deemed active, the call will be routed to the appropriate VLR for paging the mobile. This was shown in Figure 3.19 above.

Mobiles will also re-register periodically (typically every 15 minutes) to let the system know that they are still active. If unsuccessful, they will be marked inactive by the system after a period of 5 minutes and thus not paged. This is the process that occurs when a mobile is switched off, or is temporarily unable to register due to loss of signal.

Mobiles also re-register when they cross from one switch area to another, resulting in a cancellation of the first VLR entry and the creation of a second VLR entry on the switch serving the new area. In this way calls can be correctly routed to the mobile as it moves from one location area to another.

Subscribers, in the main, are distributed evenly across switch databases and therefore the network is dimensioned in recognition of the fact that all subscribers are roaming most of the time.

The need to provide national coverage usually results initially in a number of switches being strategically placed across the country of concern at centres of traffic in order to optimise the cost of cell-to-switch transmission links. As traffic increases, additional switch sites are acquired, to accommodate the mobile switching centres (23 are currently in use in the UK Vodafone network). These switches are distributed across 14 sites.

Again, initially, switches were fully meshed. However, in order to create a manageable network and minimise link costs, as the number of switches has increased, a two-tier approach has been adopted by the creation of an overlay *transit switching centre* (TSC) network, as used in the PSTN. MSCs are connected to at least two TSCs for security, the TSCs being fully meshed. Figure 1.8 of the introductory chapter indicated this arrangement, but left out the full connectivity in the interests of clarity.

A mobile switching network is often considered to be the implementation of an intelligent network. This is because it has to manage the mobility of the subscriber by routeing calls to him correctly as his location changes and deal intelligently with mobiles that are out of range.

4.3   Radio link signalling details

As mentioned earlier, when a mobile is switched on, it reads its MIN code. It then scans the preferred dedicated control channels. If the dedicated control channel is not set up for combined paging and access, the mobile will be told to tune to a paging channel. The mobile will then remain on this channel in the monitoring mode.

It is helpful to recall Figure 1.14 again, because of the channel nomenclature specific to TACS (and AMPS) cellular, namely the forward and reverse channel designations.

The information sent by a base station on the forward path of the dedicated control channel (FCC) is now described in detail.

4.3.1   Forward control channel messages

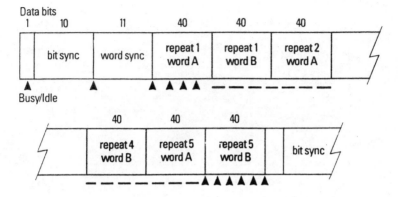

Each frame of the data stream contains bit sync and word sync for mobiles to obtain synchronisation. Busy/idle bits are sent at the beginning of every bit sync sequence, word sync sequence, first repeat of word A and every ten message bits thereafter to indicate the state of the reverse channel. The information is sent in 40-bit words and can take the form of one of three types of message:

Overhead messages
Mobile station control messages
Control filler messages

Unlike the other control and two voice channels, the forward control channel consists of a *continuously* transmitted data stream; in addition, every word is repeated five times to give adequate error protection against fading.

4.3.2   Overhead messages

These contain general data on the local system for all mobiles to receive.

Overhead messages provide the mobiles with information on the local system, such as the area identification (AI) and what access and paging channels are available in that particular cell.

The first two bits (T1, T2) will always be set to '11' to signify an overhead message.

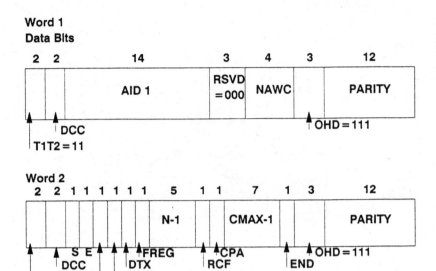

The interpretation of the data field is as follows:

TI T2   Type field. Set to '11', indicates an overhead word.

DCC   Digital colour code field; see Figure 1.5.

AI 1   First part of the traffic area identification field.

RSVD   Reserved for future use, all bits set as indicated.

NAWC   Number of additional words coming field. In word 1 this field is set to
          one fewer than the total number of words in the overhead message train.

OHD   Overhead message type field. The OHD field of word 1 is set to '110'
          indicating the first word of the system parameter overhead message.
          The OHD field of word 2 is set to '111' indicating the second word of
          the system parameter overhead message.

P   Parity field.

S   Serial number field.

E   Extended address field.

REGH   Registration field for mobile stations operating on their preferred system.

REGR   Registration field for mobile stations not operating on their preferred system.

DTX   Discontinuous transmission field.

FREG   Forced registration field.

N   The number of paging channels in the system.

RCF   Read control filler field.

CPA   Combined paging/access field.

CMAX   This is the number of access channels in the system.

END   End indication field. Set to '1' to indicate the last word of the overhead message train; set to '0' if not last word.

4.3.3   Mobile station control messages

These are specific to a particular mobile, and contain the following messages:

Page mobile	=	MIN
Power level	=	VMAC
Voice channel	=	CHAN
SAT frequency	=	SCC

Mobile station control messages are sent to tell individual mobiles what is required of them. Messages for even-numbered mobiles are sent in word A, odd numbers in word B, shown in the bit sequence of FCC message.

The message may be up to four words long and will have the first two bits (T1, T2) set to '00' if a single word is sent, otherwise multiple word messages will have word 1 set to '01', with remaining words set to '10'.

The message will always contain the mobile identity number (MIN). Depending on the action required, there may also be the mobile *attenuation code* (VMAC), the SAT colour code (SCC) and the voice channel (CHAN) assigned.

4.3.4   Control filler messages

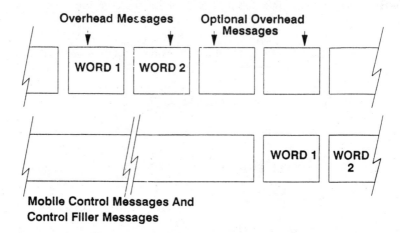

Mobile Control Messages And
Control Filler Messages

Control filler messages are data words sent to ensure the continuous stream of data on the forward control channel; in other words they can be monitored as a continuous carrier, and contain certain additional data fields indicating whether the overhead message must be read before attempting a system access.

These messages also contain power level information for the mobile on the reverse control channel and the digital colour code.

4.4   Registration

Registration is used by mobiles to announce their current location and enable the network to direct incoming calls to the appropriate cells. Mobiles are forced to register when crossing from one traffic area to another, or on command from a base station on a periodic basis, as depicted in Figure 3.18.

The mobile will have powered up, stored information from a dedicated control channel and then gone into the monitoring mode, where it listens to control channel messages.

If it is necessary to perform a registration the mobile must first access the system. This means monitoring the busy/idle bits on the forward control channel and attempting to seize the channel when it is idle. On seizing the channel, it sends a burst of identification data on the reverse control channel.

4.4.1   Reverse control channel messages

The reverse control channel (RCC) is used by the mobile to send information to the network. It is sent as a burst of data and, like the FCC, each word is repeated five times, as shown on the next page.

Bit sync and a coded DCC are sent before the information words. The two-bit DCC is derived from a seven-bit DCC sent by the base station to ensure that the correct base station has been seized.

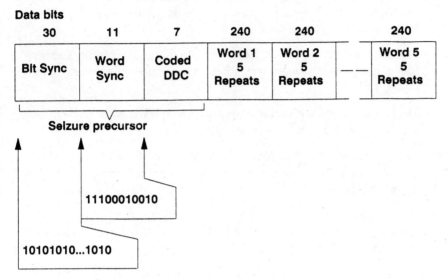

If the mobile is only performing a registration it will send three words containing the mobile identity number (MIN), electronic serial number (ESN) and other data such as *station class mark* (SCM) type of mobile.

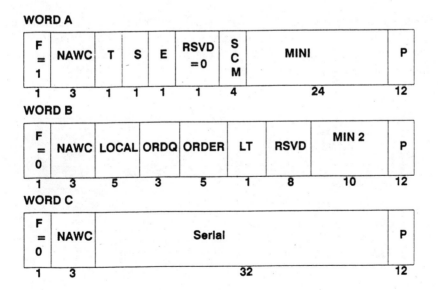

4.5 Mobile call initiation

• The mobile user starts the process by selecting a number to be called; this is normally done by loading a number via the key pad. When the *send* key is pressed the mobile checks the system access data, monitors the busy/idle bits and performs a system access.

• The mobile sends data over the RCC containing the mobile identity number, the ESN and the called number. The process is now close to the scheme shown in Figure 4.1.

• The system changes the busy/idle bits to busy and processes the received mobile data, checking that it is valid on the system.

• The system sends a mobile control message to allocate a voice channel for the conversation and at the same time sets up the call on the voice channel, sending the relevant SAT tone.

• The mobile checks the data and stores it in memory, then moves to the voice channel where it transponds the SAT frequency (one of three) to confirm that the channel is set up.

• A conversation path is now open and the mobile user will hear a ringing tone until the call is answered.

4.5.1 Reverse control channel formatting

Up to four additional words will follow the previous three words shown above on the RCC, if a call is initiated. The additional words will contain the dialled digits. TACS originally allocated two words for up to 18 dialled digits; however, later mobiles have the facility to send up to 32 dialled digits, i.e. 4 words.

Word D

F = 0	NAWC	1st Digit	2nd Digit					7th Digit	8th Digit	P
1	3	4	4	4	4	4	4	4	4	12

Word E

F = 0	NAWC = 000	9th Digit	10th Digit					15th Digit	16th Digit	P
1	3	4	4	4	4	4	4	4	4	12

4.6 Mobile call reception

- On receipt of an incoming call the system generates a mobile station control message over the FCC to page the mobile.

- The mobile monitors the busy/idle bits, and when the control channel is free, performs a system access by sending data over the RCC containing its MIN, ESN and a paging order confirmation message.

- In response to the paging order confirmation message, the system changes the busy/idle bits to busy and processes the mobile area to check that it is valid on the system.

- The system sends a mobile station control message to allocate a voice channel for the conversation and at the same time sets up the call on that voice channel, sending the relevant SAT tone.

- The mobile checks the data and stores it in memory, retunes to the voice channel, where it transponds the SAT to confirm that the channel is set up.

- The system sends an alert order to the mobile over the forward voice channel; the format is shown here.

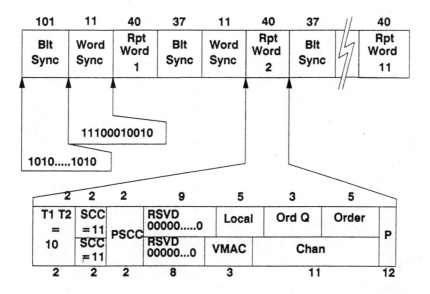

- The mobile alert device is activated and the mobile confirms this by sending an 8 kHz signalling tone (ST) over the reverse voice channel.

- When the call is answered, the signalling tone is removed.

- Removal of the ST allows a conversation path to be opened and conversation to take place.

Once a call has been set up, data will only be sent over the FVC to change power levels, perform handovers and send additional service request information.

The forward voice channel message contains a single 40-bit word repeated 11 times between syncs and word syncs.

The reverse voice channel can contain up to five words consisting of either, order confirmation for the base station, or a called address for an additional service request, as shown here.

Data bits

101	11	48	37	11	48	37	
bit sync	word sync	repeat 1 word 1	bit sync	word sync	repeat 2 word 1	bit sync	

48	37	11	48		11	48
repeat 5 word 1	bit sync	word sync	repeat 1 word 2		word sync	repeat 5 word 2

All messages contain bits to determine the number of additional words, type of word and parity.

In summary, the data format on the four channels connecting each BS and MS, are as follows:

	Bits per word	No of repeats	No of words
FCC	40	5	Continuous
RCC	48	5	3-7
FVC	40	11	1
RVC	48	5	1-5

4.7 The signalling tone

In addition to the data, the two supervisory audio tones, used for additional control via the voice channels, are very important; one is the signalling tone (ST).

ST is a tone of 8 kHz ±1 kHz modulated at a nominal deviation of 6.4 kHz and is used for four activities:

- *Confirmation of handover request:* on receipt of a command to 'handover' the mobile stores the new channel number, SAT, and power level, and sends ST for 50 ms before handing over.

- *Hookflash:* during conversation for additional services. The user loads the type of service via his keypad and presses the 'send' key. A 400 ms burst of ST is sent over the RVC to request a hookflash.

- *Cleardown:* on termination of a call by the mobile user, ST is sent for 1.8 seconds over the RVC.

- *Confirmation of alert:* after a mobile is alerted, ST is sent via the RVC until the call is answered by the mobile user.

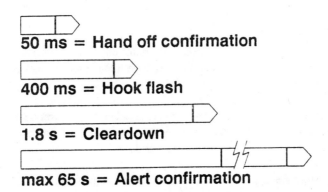

50 ms = Hand off confirmation

400 ms = Hook flash

1.8 s = Cleardown

max 65 s = Alert confirmation

4.8 The supervisory audio tone

The second tone is the supervisory audio tone (SAT). These (three) tones are again a critical part of analog cellular and have two functions.

As described in Chapter 1, section 9, it enables base stations in one particular cluster to be distinguished from base stations in a neighbouring cluster using the same channel frequencies. The MS is therefore constrained to operate to the base stations in its vicinity even though it could be receiving an adjacent cluster channel frequency, but with the incorrect SAT.

The second use of the SAT is to maintain a closed identification loop for the base station. If SAT is lost during a call the mobile unit starts a timer and, if the SAT is not received before the timer expires, the call is terminated. The mobile is advised which SAT to expect at initial call set up by the SAT colour code (SCC) on the FVC, shown in the list below.

The frequency of the generated SAT should be accurate to 1Hz.

SAT Frequency (Hz)	SCC
5970	00
6000	01
6030	10

4.9   Handover

- If during conversation the received signal strength falls low, as determined by a SINAD measurement at the BS receiver, and the mobile is on the maximum power level for the particular cell, the base station sends a message to the mobile switching centre.

- The system, by means of the BS-MSC heirarchy, initiates a search for a better cell by requesting adjacent cells to measure the signal strength of the relevant mobile.

- If a stronger signal is found, and a free channel on a (new) frequency is available in that cell, a second voice path is set up through that cell and bridged across to the existing one in preparation for a handover.

- The system generates a handover order over the initial forward voice channel.

- The mobile stores the handover data which includes the new channel number, SAT and power level; it then sends a signalling tone ST for 50 ms and turns off the initial reverse voice channel.

- The mobile re-tunes to the new voice channel, turns its transmitter on, and the new SAT is transponded.

- When the system detects the SAT, the former base station channel is released for possible other activity.

4.9.1 Illustration of signalling procedures

It is helpful to visualise the signalling procedure of cellular as calls are set up by means of a 'moving' diagram.

Figure 4.4 shows the procedure which takes place for making a call from the mobile subscriber to a subscriber in the fixed network (or it could be another mobile located in the network).

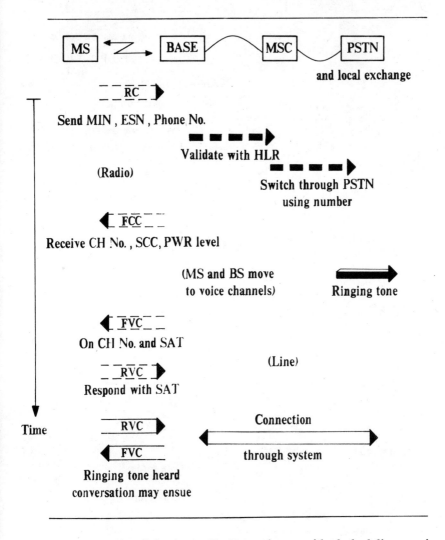

Figure 4.4   Mobile-originating call. (Data shown with dashed lines; voice signals shown as continuous lines)

The local MSC may contain the HLR, but certainly a VLR. The data transmission process on TACS requires a minimum of 2620 bits (because of the intensive repetition) which takes 327 msec at an 8 kbps signalling rate. In AMPS the signalling pattern is similar, except that it occurs at 10 kbps (in a 30 kHz BW). In NMT the signalling rate is only 1200 bps, but only requires 64 signalling bits.

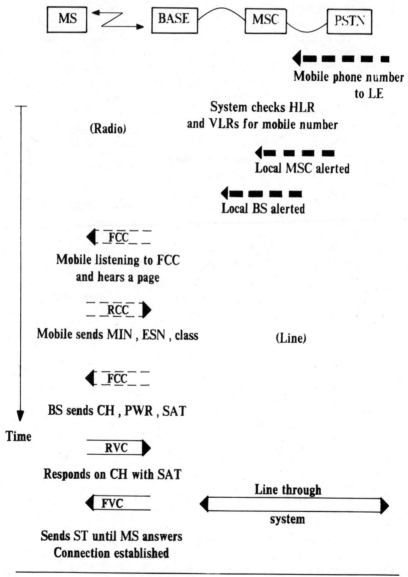

Figure 4.5   Mobile-terminating call

The action for a fixed network subscriber calling a mobile is shown in Figure 4.5.

In the case of handover, data signalling occurs only over the forward and reverse voice channels. These are above the voice (audio) band and the user is unaware of the signalling, except for a short muting action, usually of under 300 msec duration. The action is shown in Figure 4.6.

It is worth noting that extensive signalling takes place in cellular radio, chiefly because of the roaming characteristics of the mobile subscribers. However, a very large proportion of the data bits used are for signal security, either by extensive code correction or multiple repetition. This is due to the unavoidable multipath propagation condition in UHF radio circuits. We explore multipath propagation and its effect on signalling in Chapter 5.

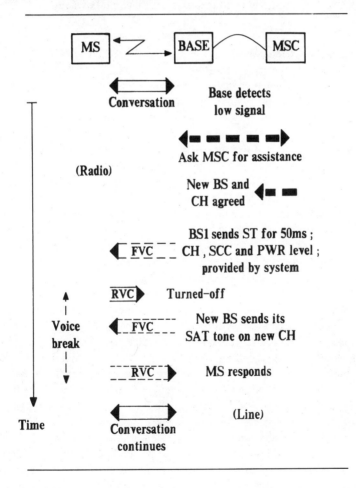

Figure 4.6   Handover activity

4.10   Data over cellular

The interesting fact about signalling within an analog cellular radio network is that much of the messaging is digital. There is therefore good reason to suppose that data messaging services can be set up on an analog cellular system - and even more reason to suppose data services can be expected within a digital cellular network. However, there are some data specific services which make definite use of the cellular radio principle. By these one does not mean data-over-voice services with mobile terminals, or even facsimile. Such services use the voice channels in analog cellular, as set up by the digital signalling protocol just described. There are of course several difficulties with this approach, due to marginal error rates on a cellular voice path, and ARQ as well as FEC for error correction are used. The methods and systems are beyond the scope of this text, however.

4.10.1 Data specific networks

It is of interest on the other hand to describe one specific network because it illustrates how a different system, or perhaps service, can come about due to the differences between cellular systems, rather than principle, as set out in Chapter 1, section 12. Such a particular system is known as PAKNET and provides a packet data network, by means of transparent VHF cellular radio operation.

The basic elements of the scheme are shown in Figure 4.7. A similar set is distributed over a seven-cell cluster like Figure 3.11; the main difference being that the cell radii are generally larger because lower VHF frequencies are licensed for the service. The precise frequencies are indicated in Table 4.1.

Fourteen channels are allocated to the base stations which contain the *network access controller* (NAC); fourteen lower frequencies are allocated to the receivingstations containing a *network terminating unit* (NTU). Only seven pairs with 25 kHz channel spacing are used; the extra pairs are for standby or possible system expansion. Note that the T_x/R_x spacing is now only 4.5 MHz, unlike the 45 MHz of conventional cellular. Also note that the service range can be some three times greater compared to conventional cellular, a point illustrated by Figure 2.11.

Because PAKNET is a data specific network, the control channel(s) in each cell acts as both control and signalling. Thus, whereas cellular signals over the forward and reverse control channels, it has to switch to voice channels for messaging, as shown in Figure 4.1. In PAKNET only two channels are involved, signalling in the time slots shown in Figure 4.8. The signalling rate used is 8 kbps, using direct frequency modulation of each RF carrier. The protocol used is a random multiple access protocol, known as dynamic slotted reservation, or ALOHA. Thus, whereas the access periods in Figure 4.8 are marked by the base station, the request-to-transmit a data packet by an NTU, is random among a set of users in a particular cell. This is what leads to the maximum number of users that one particular cell can

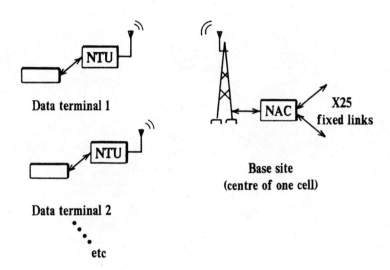

Figure 4.7   The principal components within the PAKNET data specific network

Table 4.1   The frequencies allocated to PAKNET

Channel	BS transmit	BS receive	
1	164.2125 MHz	159.7125 MHz	(Guard channel)
2	164.225	159.725	
3	164.2375	159.7375	
4	164.245	159.75	
— — —	· — — —	— — — —	
13	164.3625	159.8625	
14	164.375	159.875	
15	164.3875	159.8875	
16	164.4	159.9	(Guard channel)

(Compare this Table to Figure 3.1, for example)

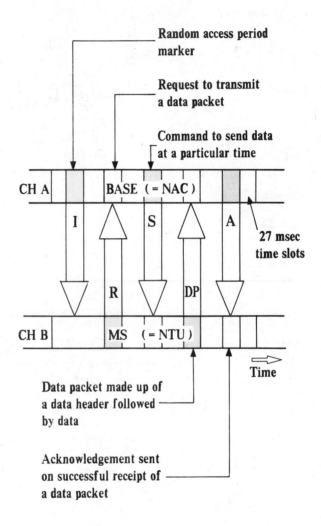

Random access period
marker

Request to transmit
a data packet

Command to send data
at a particular time

27 msec
time slots

Data packet made up of
a data header followed
by data

Acknowledgement sent
on successful receipt of
a data packet

Figure 4.8   Signalling (packet) activity in the forward and return channels of the
PAKNET system

support, because of possible collision in the request-to-transmit period R. However, the system has been dimensioned so that up to 800 terminals, operating at an average rate of one transaction (packet) every 15 minutes, can be supported in each cell.

To protect the user data, typically 128 bytes, a 12:12 FEC protocol is used, backed up by a 16-bit *cyclic redundancy check* (CRC) to detect any remaining errors. Since the system in principle is used for monetary transaction data services, no transmission errors can be accepted. Hence if the CRC identifies any errors, an ARQ operation, as discussed in section 1.10, is put into operation, that is, via transmission block A in Figure 4.8. Apart from these details, the closeness of the system to the signalling part of analog cellular is very evident.

Further reading

Davie, M.C. and Smith, J.B. (1991). 'A cellular packet radio data network', *Elec & Comms J*, June, pp 137-143

Flack, M. and Gronow, M. (1990). *Cellular Communication for Data Transmission*, NCC Blackwell, UK

Hughes, C.J. and Appleby, M.S. (1985). 'Definition of a cellular mobile radio system, *Proc IEE*, 132 Part F, August, pp 416-424

Lee, W.C.Y. (1989). *Mobile Cellular Communication Systems*, McGraw-Hill, USA, Ch 3. (The description here refers to AMPS; some of the acroynyms for the same functions as TACS are slightly different, also the signalling rates and deviations differ)

Parsons, J.D. and Gardiner, J.G. (1989). *Mobile Communication Systems*, Blackie, UK

Philips Telecommunication Review (1983). 'Special issue on mobile radio', April

5 The Multipath Propagation Problem

5.1 General considerations

The properties and principles outlined so far in respect of cellular radio would appear to show that a roaming radiotelephone system can be set up quite specifically. Radio coverage in a cell can be well defined; co-channel interference from adjacent cells can likewise be defined; analog voice with FM operates in an assured manner; signalling over the network can be defined accurately. What, then, are the remaining problems, and why is there the desire to move to digital cellular radio?

There are two main problems:

(i)   The radio signals are much less well defined than we have so far indicated. Scattering and reflection, plus movement of the user, causes what is known as *multipath propagation*. This effect causes errors in the signalling - hence the large amount of redundancy in the signalling patterns - and also puts limits on the co-channel operation in closely packed cells.

        The cell packing determines the efficiency of the cellular scheme and in turn the number of users. Digital cellular technology appears to offer improved efficiency of the limited available radio spectrum, a subject discussed in some detail later towards the end of the book.

(ii)   The fixed telephone network is nearly all digital; all voice messages are encoded and transmitted in a PCM format. The same arrangement is the aim of the mobile phone network operators. All messages and signalling would be digital.

        Because the bandwidth available in the mobile (radio) part of the network is severely limited, however, more speed-efficient coding algorithms are necessary. Also, the digital signalling rate per Hertz should be high. Digital cellular has to address both these problems, again under the shadow of multipath propagation.

A very important advantage that can also arise by switching to digital operation is that of 'universality'. The network can be extended across national borders, allowing international roaming of subscribers. This concept is described in Chapter 8.

5.2 Multipath fading characteristics

The multipath problem in mobile radio is caused by reflection and scattering from buildings, trees and other obstacles along the radio path. Radio waves arrive at a mobile receiver from many different directions, with different time delays. If one refers back to Figure 2.6, an extra ray path was shown coming from a scattering source (a building). Together with a possible direct ray, a ground reflected ray and other possible scattered rays, these combine vectorially at the receiver antenna to give a resultant signal which depends on the differences in path length that exist in this multipath field. Also, as a vehicle-borne, or handheld, receiver moves from one location to another, the phase relationship between the components of the various incoming waves changes so the resultant signal changes. It is important to note that whenever relative motion exists there is a Doppler shift of the frequency components within the received signal.

Characterising the mobile radio channel is therefore not a simple task. It is possible, however, to deal with the problem in two ways. Firstly, we can consider the case where the signals occupy only a narrow bandwidth. By 'narrowband' we mean that the spread of *time delays* in the multipath environment is sufficiently small for all spectral components within the transmitted message to be affected in a similar way. By time delays we refer to the time of travel, at the speed of light, from the transmitter (BS, say) to the mobile, and these will clearly vary according to the diversity of the path. Previously, we had referred to these difference as phase differences. At a frequency f, over a path length difference of 2π, or one wavelength, the excess path delay is clearly 1/f, see Figure 2.1. At 900 MHz this time delay difference would be 1.1 nanoseconds.

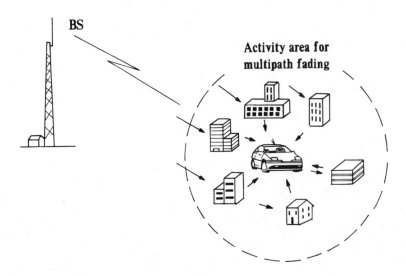

Figure 5.1   Illustration of radio propagation in an urban area

In the narrowband case there are no frequency-selective effects and the characteristics of the channel can be expressed in terms of their effect on any one component in the message - the carrier frequency is usually used. A more complicated form of characterisation is needed to deal with wideband signals; 'wideband' in this case is used to indicate that frequency-selective effects do occur.

First we deal with the narrowband case in order to introduce ideas and terminology relevant to this subject. In urban areas problems exist due to the fact that the mobile antenna is low, so there is no line-of-sight path to the base station which itself is often located in close proximity to buildings. Propagation is therefore mainly by means of scattering and multiple reflections from the surrounding obstacles, as shown in Figure 5.1. Because the wavelengths in UHF bands are less than 1 metre, the position of the antenna does not have to be changed very much to change the signal level by several tens of dB.

This feature is known as slow fading and is observed as the position of the mobile is changed. It can be experienced in FM car radios and VHF private mobile radio (PMR) equipment. The signal appears to vanish at certain positions, but moving only a few metres brings it back again. The signal envelope takes on a standing wave pattern, as shown in Figure 5.2, and is characterised by a log-normal probability density function. The mean signal level, about 3 dB below any peak value, is the one calculated in Chapter 2.

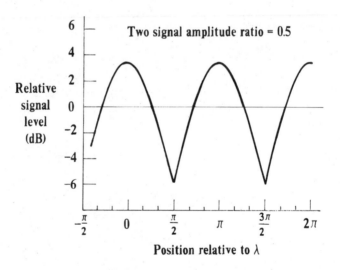

Figure 5.2   Standing wave pattern due to two carrier waves being received at the same time versus their relative phase

A receiver moving continuously through this field experiences a time-related variable signal which is further complicated by the existence of Doppler shift. The signal fluctuations caused by the local multipath are now known as *fast fading*, to distinguish them from the much more leisurely variation in mean level, referred to above as slow fading. A record of fast fading is shown in Figure 5.3. Fades of a depth less than 20 dB occur quite often, but deeper fades in excess of 30 dB are less frequent. Rapid fading is usually observed over distances of about half a wavelength; therefore, at VHF and UHF, a vehicle moving at 50 km/hr can pass through several fades in a second.

Examination of Figure 5.3 shows that it is possible to draw a distinction between the short-term multipath effects and the longer-term variations of the local mean. Here, however, we consider only the short-term effects both for narrowband and wideband channels; in other words, we consider signal statistics in which the mean level of the signal is constant.

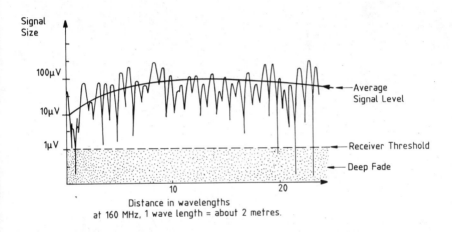

Figure 5.3   Illustration of typical envelope pattern of a VHF signal received under multipath conditions

5.2.1 Elementary multipath

A multipath propagation situation always contains several different paths by which energy travels from the transmitter to the receiver. If we consider the case of a stationary receiver, one can imagine a static multipath situation in which several versions of the signal arrive sequentially at the receiver. The effect of the differential time delays is to introduce relative phase shifts between the component waves and superposition of the different components leads to either constructive or destructive addition, depending upon the relative phases. Figure 5.4 illustrates the resultant signal arising from two paths.

When either the transmitter or the receiver is in motion, we have a dynamic multipath situation in which there is a continuous change in the electrical length of every propagation path and thus the relative phase shifts between them change as a function of spatial location.

The time variations, or dynamic changes in the propagation path lengths, can be related directly to the motion of the receiver and indirectly to the Doppler effects that arise. In a practical case, the several incoming paths will be such that their individual phases, as experienced by a moving receiver, will change continuously and randomly.

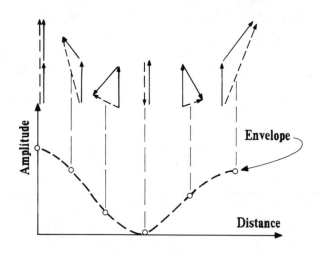

Figure 5.4   Illustrating how the envelope fades as two signals combine with different phases

An established multipath model assumes that the field incident on the mobile antenna is composed of a number of plane waves of random phase, these plane waves being vertically polarized with nearly horizontal angles of arrival and phase angles which are random and statistically independent. Further, the phase angles are assumed to have a uniform probability density function in the interval $(0,2\pi)$. This is reasonable at VHF and above, where the wavelength is sufficiently short to ensure that small changes in path length result in significant changes in the RF phase.

5.2.2   A scattering model

Continuing with this model, therefore, at every receiving point, we assume the existence of n plane waves of similar amplitude. This is a realistic assumption in heavily built-up areas, for example, since the scattered components are likely to experience similar attenuation and there will be no dominant component. In certain situations, a direct line-of-sight path may contribute a steady non-random component, but here we restrict our discussion to the case of n equal-amplitude waves.

The path-angle geometry for the scattered plane waves is shown in Figure 5.5.

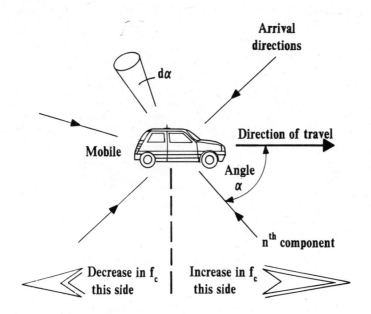

Figure 5.5   Path arrival geometry at a mobile station in a multipath situation. Vehicle motion is indicated

If the transmitted signal is vertically polarized, we assume that the components at the mobile have a vertical electric field E_z. Also, the assumption that the mobile received signal is of the scattered type, with each component wave being independent, randomly phased and having a random angle of arrival, leads to the result that the *probability density function* of the envelope R (the vector sum of the quadrature X and Y components) is

$$p_r(R) = \frac{R}{\sigma^2} \exp\ (-R^2/2\sigma^2) \qquad\qquad \ldots (5.1)$$

This is called a Rayleigh distribution. The corresponding *cumulative distribution function* P_r is

$$P_r(R) = 1 - \exp(-R^2/2\sigma^2) \qquad\qquad \ldots (5.2)$$

where σ^2 is the mean power. These two functions are shown in Figure 5.6.

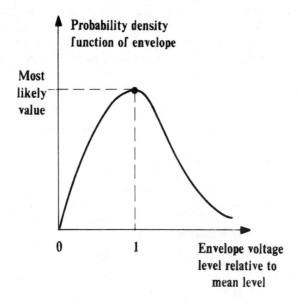

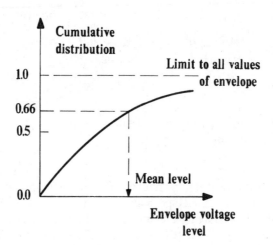

Figure 5.6   The Rayleigh probability density and cumulative distributions func-
              tions

Since one is considering the envelope of the fading signal, both distributions are
always positive functions.

A useful way of understanding the multipath signal, and why it is called
Rayleigh fading, is to note a method of simulating the signal. Thus, as shown in
Figure 5.7, an RF signal is divided into two paths, an in-phase path X, and an out-

of-phase path Y. Each signal is then modulated by a Gaussian amplitude voltage function which gives to the two components (X and Y) the required signal (±) amplitude variance. Adding the two components produces a signal with the Rayleigh envelope fading function p(R).

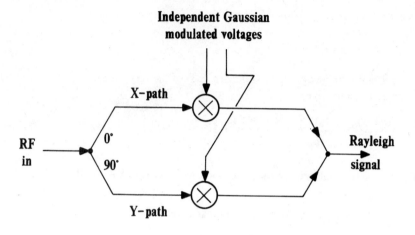

Figure 5.7   Two-path modulation approach to Rayleigh fading signal generation

5.2.3 *Effect of vehicle velocity*

If either the transmitter or receiver is in motion, the components of the received signal each experience a Doppler shift, the frequency shift being related to the angle between the direction of arrival of that component and the direction of vehicle motion. For a vehicle moving at a constant speed v along the X-axis, in Figure 5.5, the Doppler shift f_n of the plane-wave component arriving at an angle α_n, is

$$f_n = \frac{v}{\lambda}\cos a_n \qquad \qquad \ldots (5.3)$$

It can be seen that component waves arriving from ahead of the vehicle experience a positive Doppler shift (maximum value $f_d = v/\lambda$), whilst those arriving from behind the vehicle have a negative shift.

It is worth noting that a speed of 40 mph at 900 MHz, produces a maximum Doppler shift of 53 Hz, i.e.,

$$f_d = \frac{v \text{ (metres per second)}}{\lambda \text{ (metres)}}$$

$$= \frac{40 \times 8/5 \times 1000}{0.33 \times 60 \times 60} = 53 \text{ Hz}$$

using the conversion of 8 km to every 5 miles, etc. Also, a proportional change in frequency, or speed, will produce a proportional change in f_d.

If n is large, the fraction of the incident power contained within angles between α and $\alpha + d\alpha$, for an omnidirectional antenna, is $p(\alpha)d\alpha$. Equating this to the incremental power determined through the relation between the Doppler shift and the nominal carrier frequency f_c, namely,

$$f(\alpha) = f_d \cos \alpha + f_c$$

we can obtain the power spectral density S(f) of the received signal, observed by a vertical antenna, namely

$$S_{EZ}(f) = \frac{1.5}{\pi f_d} \left[\frac{1 - (f - f_c)^2}{f_d} \right]^{0.5} \qquad \ldots (5.4)$$

This power spectral density function is shown in Figure 5.8.

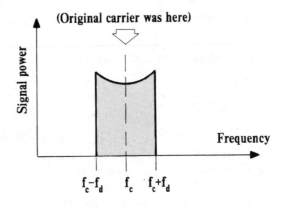

Figure 5.8   Power spectral density function of an RF carrier caused by Rayleigh multipath

If a dominant component exists in the incoming signal, then this has a substantial influence on the signal spectrum. For example, such a component arriving at an angle α_o gives rise to a spectral line at $f_c + f_d \cos \alpha_o$ in the RF spectrum.

In the time domain, the effects of the randomly phased and Doppler-shifted multipath signals appear in the form of a fading envelope, as described above.

5.2.4   Fading envelope statistics

The fading envelope directly affects the performance of any receiver. The Rayleigh fading envelope only occasionally experiences very deep fades, for example 30 dB fades occur for only 0.1% of the time. This can be understood from the signal envelope cumulative probability distribution, shown in Figure 5.9. To specify some of the constraints in a quantitative manner, one parameter in particular that interests us, is how often the envelope crosses a specified signal level.

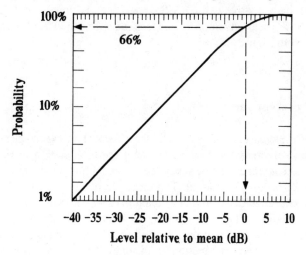

Figure 5.9   Cumulative probability distributions of the signal envelope RMS level

The level crossing rate L at a specified signal level R is defined as the average number of times per second that the signal envelope crosses the level in a positive-going direction.

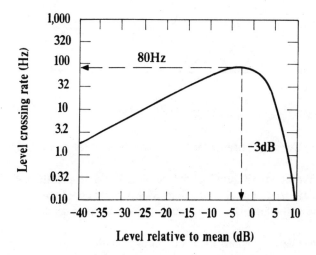

Figure 5.10   The level crossing rate of a fast fading signal envelope versus level relative to RMS value for a 80 Hz Doppler

Figure 5.10 shows a calculated result of this function. The level crossing rate L is shown in Hertz, but is really relative to the maximum Doppler shift frequency f_d. Figure 5.10 assumes that $f_d = 80$ Hz. The levels at which crossings occur are indicated as decibels relative to the RMS signal level. We note that at $L = f_d$ is a level 3 dB below the peak envelope level. Deep fades are much less likely to occur, as was indicated in Figure 5.3.

5.3 Diversity reception

In mobile radio systems, especially at UHF, the effects of fading can be reduced by the use of space diversity techniques, either at the base station or the mobile, provided that two antennas can be separated far enough to ensure that the signal envelopes exhibit a low correlation. The principle of diversity is based on the observation that the envelope covariance function of the component E_z of the signal at the mobile in Figure 5.5, and the same signal at a position removed by a spatial distance X/λ, in the case of isotropic scattering, obeys a function which is the square of the first order Bessel function (as in analog FM theory). This function is of course always positive and less than or equal to one.

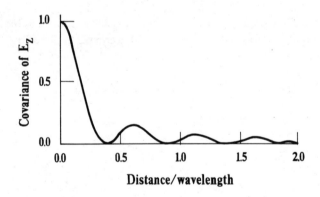

Figure 5.11   Covariance function for the envelope of the electric field

The result is plotted in Figure 5.11 for the case of an isotropically scattered field. There is rapid decorrelation, showing that space diversity could be implemented at the mobile end of the link where the assumption of isotropic scattering is approximately true.

On the other hand, it is less clear that at base station sites the assumption of isotropic scattering will hold. Base station sites are in general chosen to be well above local obstructions in order to give the best coverage of the intended service area. On the other hand the scattering objects, which produce the multipath effects, are located principally in a small area surrounding the mobile, see Figure 5.1. The reciprocity theorem applies, of course, in a linear medium, but this should not be taken to imply that the spatial correlation distance at one end of the radio path is the same as it is at the other.

It is clear that antennas have to be further apart at base station sites than at mobiles to obtain decorrelation, and indeed this is the case. It is also apparent that whereas at mobiles the assumption of isotropic scattering, from scatterers which surround the mobile uniformly, leads directly to the conclusion that the correlation between the electric field at two receiving points is a function of their separation only, this is not the case at base station sites. Here, scattering is not isotropic and the correlation between the electric field at two receiving points is a function of both their separation and the angle between the line joining them and the direction to the mobile.

5.4 Frequency selective fading

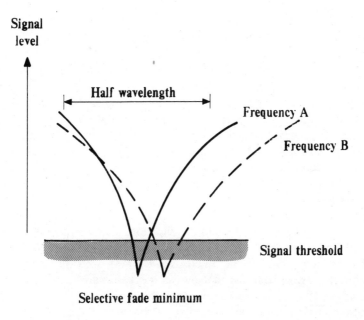

Figure 5.12   Nulls in reception due to two-path interference because of selective fading. Note how the null of frequency B does not coincide with frequency A

The previous discussion described the envelope and phase variations of the signal received at a moving vehicle when an unmodulated carrier is radiated by the base station transmitter. The question now arises as to whether this description of the channel applies when signals, which occupy a finite bandwidth, are radiated. It is clear that we need to consider the effects of multipath propagation in the case of two or more frequency components within the message bandwidth. If these frequencies are close to each other, then the different propagation paths within the multipath field have approximately the same electrical length for both components, and their amplitude and phase variations will be very similar. This is the narrowband case. As the frequency separation increases, however, the behaviour at one frequency tends to become uncorrelated with that at the other frequency, because the differential phase shifts along the various paths are quite different at the two frequencies.

An example of frequency selective fading is shown in Figure 5.12. One frequency could well suffer a severe fade, whereas the second frequency's fade is delayed in position. This illustrates the potential for diversity by frequency hopping during messaging, a technique employed in the GSM digital cellular system.

5.5   Coherence bandwidth and delay spread

The extent of the decorrelation depends on the spread of time delays, because the phase shifts arise from the excess path lengths. For large delay spreads the phases of the incoming components can vary over several radians, even if the frequency separation is quite small. Signals which occupy a bandwidth greater than that over which spectral components are affected in a similar way will become distorted, because the amplitudes and phases of the various spectral components in the received version of the signal are not the same as they were in the transmitted version. The phenomenon is known as *frequency selective fading*; the band over which the spectral components are affected in a similar way is known as the *coherence bandwidth*.

Figure 5.13 attempts to explain the situation. Here we have three possible Rayleigh power spectral density functions, as in Figure 5.8, but now delayed in time relative to each other, as the principal scatters are markedly spaced in distance, those furthest away giving rise to the greatest excess path delay.

However, all the scattering comes from the source, say the base station, and hence the signal power has to be distributed amongst the three cases. In practice, the situation is much more like Figure 5.14. The Doppler shift of every component remains the same, that is f_d or less, but the spectral envelope will be a function of the particular environment. Also one can only describe a *delay spread* s, rather than a particular delay.

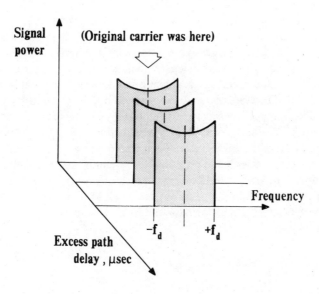

Figure 5.13   Power spectral density function showing different path delays

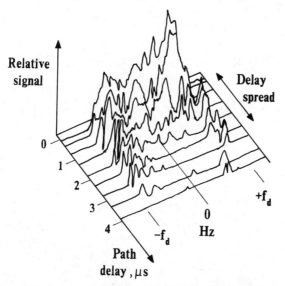

Figure 5.14   Typical scattered power profile in a suburban environment

The coherence bandwidth is inversely proportional to the delay spread. Therefore, a large delay spread (in microseconds, say) signifies a small coherence bandwidth (of the order of some 100 kHz), and manifests itself in the form of frequency selective fading for wideband type signals.

Extreme values of average delay and delay spread occur quite frequently in suburban areas; they are most common in environments with high-rise residential buildings, for example, irrespective of street orientations. Urban areas show comparable extreme values, though less frequently in the case of the delay spread. The outer fringe of the urban environment often represents the worst multipath, being influenced to a greater extent by remote tall buildings some distance away. The effects of irregular scattering are also evident in the statistical distributions for the coherence bandwidths. Variability of the parameters is greater in suburban areas, although the extreme values are comparable.

This type of large scale characterisation can be applied to the evaluation of radio systems and performance criteria emerge in terms of the percentage of locations where the performance requirements are achieved. In digital cellular radio such performance measures are laid down for multipath worst-case conditions development and testing.

Further reading

Jakes, W.C. (Ed.) (1974). *Microwave Mobile Communications*, Wiley Interscience, NY

Lee, W.C.Y. (1982). *Mobile Communications Engineering*, McGraw-Hill Book Co, NY

Parsons, J.D. (1992). *The Mobile Radio Propagation Channel*, Pentech Press, London

6 Modulation Techniques

6.1 Introduction

Modulation is the process whereby the message information is added to the radio carrier. The carrier wave was depicted in Figure 2.1. This is the carrier which attempts to get to and from the mobile, subject to the propagation losses and multipath effects described earlier. Adding modulation widens the bandwidth of the signal. Ideally, a carrier has no bandwidth, but in practice, drift, noise and propagation produce observable bandwidth. Modulation produces sidebands, either on one side, or on both sides of the nominal carrier frequency, usually having a distribution of amplitude over the assigned bandwidth. In effect one could say that whatever the modulation, the situation will exist in which one has a concentrated group of closely knit carriers, or signals, coming to and going from the mobile, *whether one specifies analog, or digital modulation.*

The question that one needs to ask, or solve, is what is the best way of arranging this group of signals; analog frequency modulation (FM) or continuous phase minimum shift keying (MSK), for example? Up until the present time, narrowband FM has dominated the cellular market.

However, recently the world has been moving to digital operation and the realisation that the control and switching between users can be performed much better by digital operation. Why not use digital modulation throughout the system? Can the signalling group around the carrier be arranged as efficiently as in the case of FM? What one finds is the converging of all forms of modulation to form the transmitted signal package; the difference between analog and digital modulation to some extent being the more extensive use of digital signal processing prior to modulation and often after demodulation in the digital case.

6.2 The bandwidth problem

The bandwidth problem arises because, either one divides the subscriber channels into a well defined frequency division multiple access (FDMA) arrangement, or one divides the users into time division multiple access (TDMA) time slots, which requires that they transmit their information, promptly, at well defined intervals within the total assigned bandwidth.

In an FDMA system spectrum partitioning is arranged and, as shown in Figure 3.1, our user's carrier assignment and cluster of modulation sidebands need to be accurately placed in the assigned channel. Sidebands at the edge of the assigned channel are required to be at least 60 dB below the unmodulated carrier amplitude. This situation applies whether the modulation system is analog or digital. For example, if the system allocation is on a 25 kHz channel basis, it is clear that whatever the modulation the actual *message bandwidth* Δf_m needs to be less than the *channel assignment bandwidth* Δf_a; for example, if $\Delta f_a = 25$ kHz, then Δf_m is usually equal to 13 kHz.

To some extent it could be suggested much of the radio spectrum is being wasted because of channelization. A way around the problem is of course to assign channels on a staggered space-dispersed basis. In TDMA the users are grouped in fewer wider channels and the waste problem is reduced. However, power control, delayed multipath and near-far problems can arise.

In digital modulation a reduced bit rate per Hertz gives the possibility of having more data (user) channels. Unfortunately, for speech this is not easily achieved in practice, as described in Chapter 7, and the necessary channel bandwidth remains a definite problem.

In Chapter 9 a simple guide formula for the number of radio subscribers, in a province or city, who can expect to be offered a service at any particular time, is derived, using the variables:

Area of the city being considered	=	A km²
Population of the city	=	P thousands of people
Average radius of a radio cell	=	R km
Number of radio channels in each cell	=	n_c

n_c will depend on how the cell repeat pattern is organised and also on the spectrum allocated to the service. Chapter 3 and, in particular Table 3.1, will be useful; channel allocation is discussed there.

It turns out that the percentage of the population in a city is given by the equation

$$\% = \frac{An_c}{PR^2} \qquad \qquad \ldots (6.1)$$

The result can be rewritten involving the total bandwidth ΔF allocated to the service, the bandwidth assigned or allocated to each user Δf_a, and cell cluster size N, since

$$n_c = \frac{\Delta F}{\Delta f_a N} \qquad \qquad \ldots (6.2)$$

Hence

$$\% = \frac{A\,\Delta F}{PR^2 \Delta f_a N} \qquad \qquad \ldots (6.3)$$

This equation shows that, apart from the global terms, the success of a cellular arrangement depends on the spectrum allocated per user, that is, the term Δf_a, which as we saw above is greater than the signal bandwidth Δf_m.

Since the above equation, as explained later in Chapter 9, is an approximation, because of the trunking gain factor, population distribution factor, and also Δf_m is a percentage of Δf_a, usually only $\approx 50\%$, this implies that we can use Δf_m (the user's signal modulation bandwidth) as the benchmark of any system design.

This question of spectrum efficiency in a cell-structured radio system is discussed in detail later, also considering co-channel interference which, as discussed in section 3.4.2, is a problem centred around the cell layout geometry (N) and the co-channel interference threshold. 25 kHz analog FM is good in this respect, because it achieves a significant demodulation gain, but digital modulations are found to possess lower thresholds and smaller cell clusters are possible.

6.3 Analog modulation bandwidths

Conventional frequency modulation is the mainstay of present-day mobile radio and cellular systems; it is an established technology, and certainly 12.5 kHz FM compactly occupies the signal bandwidth Δf_m for the transmission of speech, although 25 kHz FM has noticeably superior sound quality, for example, as on TACS cellular. For completeness we repeat some well-known factors associated with FM.

The FM signal waveform can be written

$$s(t)_{FM} \quad = \quad \cos(\omega_c + \beta m'(t))t \qquad \qquad \ldots (6.4)$$

where $\quad \beta \quad = \quad$ *modulation index* of analog FM

$$= \quad \frac{\text{maximum frequency deviation}}{\text{modulation frequency}}$$

$$= \quad \frac{\Delta f}{f_m} \qquad \qquad \ldots (6.5)$$

To generate FM it is necessary to cause frequency deviation of the transmitter output frequency, this being the so-called *instantaneous carrier frequency* f_i. In order to have a stable transmitter oscillator, it is essential to minimise its deviation, but if the oscillator is only a submultiple of the final output frequency, multiplication of the carrier can be used to generate the desired frequency and correct FM index, more accurately.

Figure 6.1 shows how the carrier waveform changes in frequency, but is transmitted with a constant amplitude. Note also how the FM has a non-regular zero crossing, meaning that the carrier changes phase over the modulation cycle.

Cellular Radio

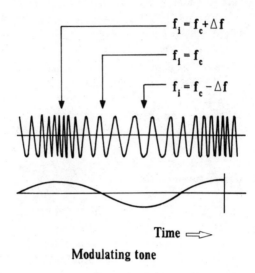

Figure 6.1   The instantaneous frequency of an FM waveform in relation to
              sinusoidal modulation

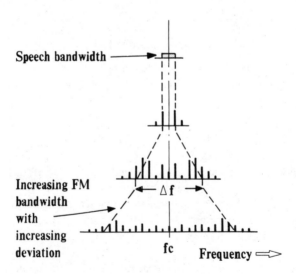

Figure 6.2   The corresponding spectrum of an FM waveform showing how the
              number of sidebands (equal spacing) increases with the modulation
              index

The spectrum shown in Figure 6.2 widens as the deviation (Δf) increases and the FM package can be viewed as a group of symmetrically packed sidebands. The operation with a voice message is such that the FM signal spectrum appears to occupy the same bandwidth for all speech, depending on the level at the microphone. To help in this matter a considerable amount of speech processing is carried out ahead of the modulation; namely clipping, filtering and pre-emphasis. A typical FM audio circuit arrangement is shown in Figure 6.3.

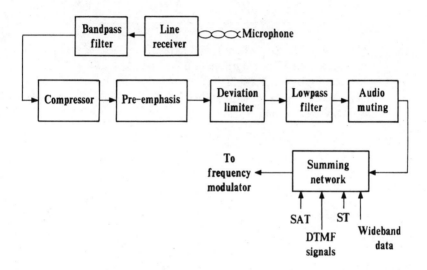

Figure 6.3   Schematic diagram of a typical cellular radio transmitter audio processing arrangement

The transmitter deviation is set to some maximum value, dictated by the agreed channel bandwidth. Table 6.1 shows three cases; the values being calculated from (6.5) and the notes appended to the table.

Apart from the bandwidth occupied by the FM waveform, column (4), the other interesting result is the improvement in the received signal-to-noise ratio on reception (and de-emphasis). Large index and hence wideband FM is clearly superior, but this result does not come without the penalty of worsening threshold. Figure 6.4 helps here. The improvement is only realised beyond a minimum threshold C/N ratio. Calling C/N the carrier-to-interference ratio C_i in cellular, one can now see why TACS and AMPS are limited to seven-cell clusters (see Figure 3.13). Very narrowband NMT and NAMPS systems are being considered and they may tolerate four-cell clusters and hence serve more users, but the subscriber will have to accept a lower voice quality.

Table 6.1   Low index FM performance data

FM index	f kHz (1)	No. of siebands (2)	Carson BW kHz (3)	S/N improvement at receiver (4)
0.8	2.4	4	10.8	+ 5 dB
1.5	4.5	6	15.0	+ 11 dB
5.0	15.0	12	36.0	+ 21 dB

Notes   (1) Using $\Delta f = \beta f_m$ where $f_m = 3$ kHz
        (2) Calculated from significant BW
        (3) Using BW $= 2\Delta f + 2f_m$
        (4) According to demodulation improvement and pre-emphasis text-
             book equations

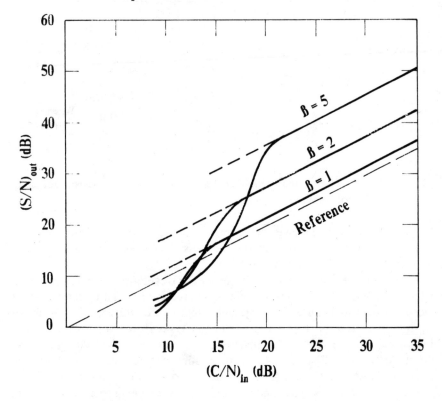

Figure 6.4   FM S/N output performance versus input C/N for different
               index modulation

An alternative analog modulation is double-sideband reduced, or suppressed carrier amplitude modulation (DSB-SC), which is an example of a modulation where the signal bandwidth fully fills the available bandwidth. Its signal-to-noise performance is as good as, if not better than 1.5 index FM, yet it only occupies less than 6 kHz bandwidth. It is referred to here because the principle of generation leads directly into digital modulations.

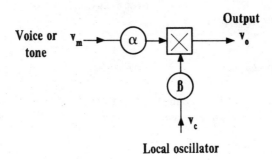

Figure 6.5   Generation of DSB-SC using a balanced mixer and showing dc offset components

Thus, Figure 6.5 shows how DSB-SC is generated using a double-balanced modulator, or multiplier. In practice, complete voltage balance of the signal input ports is not achieved, and the equation of operation of a practical modulator, including the dc offset components α and β, should be written as

$$v_o(t) = v_m(t) \cdot v_c(t)$$

$$= (v_m \sin \omega_m t + \alpha) \cdot (v_c \sin \omega_c t + \beta)$$

$$= v_m.v_c \sin\omega_m t \sin \omega_c t + \alpha.\beta + \alpha.v_c \sin\omega_c t$$

$$+ \beta.v_m \cos\omega_m t \qquad \qquad \ldots (6.6)$$

The desired output is the first product. Mixer imbalance and any non-linearity will cause difficulty with practical modulator realisation, especially when the outputs of two quadrature mixers are added together, as is done below.

The waveform of DSB-SC is shown in Figure 6.6(e) below when describing digital phase shift keying modulation. The fully filtered PSK waveform will have sidebands as single pairs if a continuous on/off waveform is the modulation.

6.4   Shift key modulations

Digital modulation leads to a family of what are known as shift key modulations. Let us assume that the binary data is coded in some suitable form, such that

a '1' or mark waveform = +1

a '0' or space waveform = −1

over the symbol period T_b

6.4.1   Phase shift keying

Applying this signal to a balanced modulator of Figure 6.5, with the amendments shown in Figure 6.6, leads to binary phase shift keying (BPSK), or bipolar ASK, that is

$$s(t) = \pm \cos \omega_c t \qquad \qquad \ldots (6.7)$$

The output waveform shown in Figure 6.6(d) has a constant amplitude, but a phase change of 180° at each mark/space transition. The spectrum of this waveform extends well beyond the symbol rate frequency f_r, offset either side from the (suppressed) centre carrier, f_c, where

$$f_r = 1/T_b$$

Note that we will now be using symbol rate, as opposed to bit rate, in order to accommodate multilevel modulations discussed below, but for binary modulation, either can be used.

To contain the spectrum, shaping or filtering of the data symbols has to be employed, see Figure 6.6.

The ideal Nyquist signalling rate in a bandpass channel filter of bandwidth B is

$$f_r = B$$

Vestigial shaping of the channel filter would most likely be employed, which reduces f_r to the value

$$f_r = \frac{B}{1 + \alpha} \qquad \qquad \ldots (6.8)$$

where $0 < \alpha < 1$

(ideal)   (raised cosine)

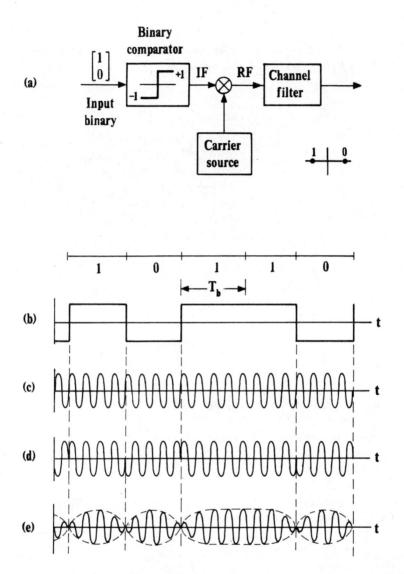

Figure 6.6   Generation of the binary phase shift keyed signal, (a) modulator, (b) data waveform, (c) carrier, (d) unfiltered BPSK, (e) filtered BPSK

A change in the filtered envelope amplitude of the signal, that is Figure 6.6(e), is seen to occur at the mark/space transition, and therefore more complex quadrature modulation is used, as described below.

6.4.2   Frequency shift keying

Frequency shift keying (FSK) implies that one switches from a mark frequency to a space frequency, in sympathy with the data, see Figure 6.7.

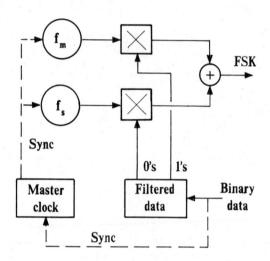

Figure 6.7   Schematic for generating FSK

The spectrum of the FSK waveform

$$s(t) = \cos(\omega_c \pm \Delta\omega)t \qquad \qquad \ldots (6.9)$$

can be calculated in the case of a regular 1,0,1,0 pattern, by adding the spectrum of the symbol rate-time limited carrier waveforms

$$f_m = f_c - \Delta f \quad \text{for period } T_b$$

$$f_s = f_c + \Delta f \quad \text{for period } T_b$$

One finds that the signal energy concentrates at the mark and space frequencies, as indicated in Figure 6.8.

The modulation index of FSK is given by

$$m = (f_s - f_m).T = \frac{f_s - f_m}{f_r} \qquad \qquad \ldots (6.10)$$

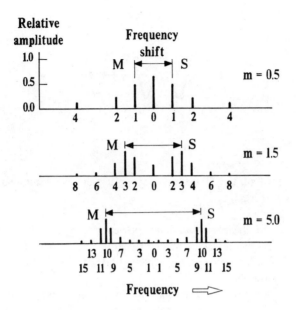

Figure 6.8   FSK signal spectrum with increasing deviation

To increase the data transmission rate, but not the bandwidth occupied, low index FSK schemes must be employed. A common standard is *fast frequency shift keying* (FFSK) where

$$f_m = 1200 \text{ Hz} \qquad \text{(note the nominal carrier}$$
$$f_s = 1800 \text{ Hz} \qquad \text{frequency } f_c = 1500 \text{ Hz)}$$
$$f_b = 1200 \text{ bps, = symbol rate frequency} = f_r$$

$$\therefore m = \frac{1800 - 1200}{1200} = 0.5$$

Here one cycle of 1200 Hz is followed by one and a half cycles of 1800 Hz, with no phase discontinuity at the bit interval. The spectrum is now concentrated in the band 600 to 2400 Hz.

It can be shown that FFSK is equivalent to *minimum shift keying* (MSK), a particular form of *quadrature phase shift keying* (QPSK), which is described below. The waveform of this modulation is sketched in Figure 6.9. It is useful to refer to this diagram when reading about MSK and the phasor diagram in Figure 6.13.

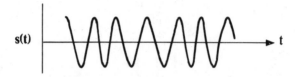

Figure 6.9   Detail of the fast frequency shift (and minimum shift) keying waveform

With all three of these modulations, however, the modulation signal has significant adjacent channel sideband components, which make them unsuitable for future digital cellular radio systems where good adjacent channel performance is an essential requirement.

The solution to the difficulty is to shape the binary signal waveform by suitable wave filtering. In the case of FSK, the conventional method is to constrain the changeover from a mark to a space, with a filter in the data source, as in Figure 6.7.

The coherence between the mark and space tone is retained by arranging that

$$f_m = f_c - nf_b/4 \qquad \text{where } n = \text{an integer}$$

$$f_s = f_c + nf_b/4 \qquad\qquad \dots (6.11)$$

where

$$f_b = \text{bit rate}$$

When $n = 1$, we have FFSK as just described. When $n = 2$, we have again, coherent FSK, with $m = 1$. FFSK can only be demodulated with a coherent demodulator; FSK with $m = 1$ can be detected with a non-coherent demodulator, but there is a penalty to pay with regard to the BER performance. However, the mark/space coherence achieved by observing the rules in (6.11), which defines *coherent frequency exchange keying* (CFEK), provides superior adjacent channel performance over and above simple FSK, as shown in Figure 6.10.

The data throughput can be increased by using more than two frequencies. Thus using L levels, or frequencies, a symbol can represent $\log_2 L$ bits; that is, four frequencies will convey two bits per symbol. This technology is being explored for a new European radio messaging (paging) system (ERMES).

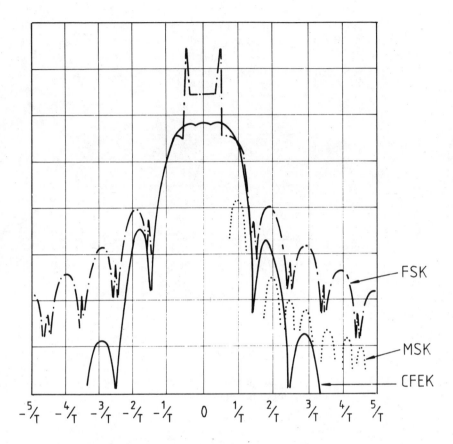

Figure 6.10   The adjacent and in-band spectrum occupancy non-coherent of FSK (m = 1), MSK (m = 0.5) and CFEK (m = 1)

The data rate planned is 6.25 kbps; it thus only requires a symbol rate of 3.125 kbps (half), and uses the four signalling frequencies

$$
\begin{aligned}
f_{00} &= f_c - 3f_b/2; & \text{offset} &= -4687.5 \text{ Hz} \\
f_{10} &= f_c - f_b/2; & \text{offset} &= -1562.5 \text{ Hz} \\
f_{11} &= f_c + f_b/2; & \text{offset} &= +1562.5 \text{ Hz} \\
f_{01} &= f_c + 3f_b/2; & \text{offset} &= +4687.5 \text{ Hz}
\end{aligned}
$$

The arrangement is a four-level equivalent of the two-level CFEK, with m = 1.0; at least between adjacent levels.

The data is shaped using a 10th order Bessel filter, with a 3 dB point at 4 kHz, so that the carrier oscillator can move smoothly from one signalling frequency to

the next within the symbol period, that is, a frequency change of 1/3.125 msec, as
sketched in Figure 6.11. It is claimed that adjacent channel interference levels of
better than −70 dB can be achieved and a modulation efficiency of approximately
1.3 bps/Hz is offered. A three-level, conventional limiter-discriminator detector
can be used, provided that the signal receiving conditions are good.

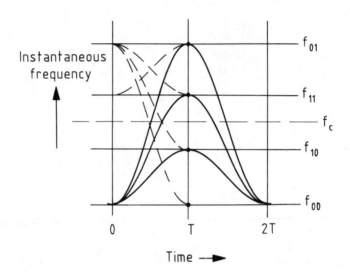

Figure 6.11   The possible instantaneous frequency positions of 4-level FSK
               over two symbol periods

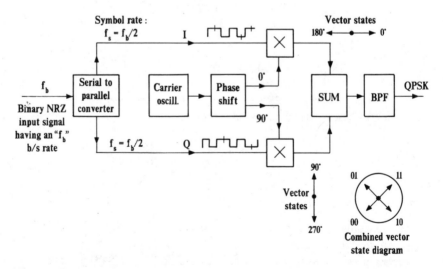

Figure 6.12   The two-path method of generating QPSK

6.4.3 Quadrature phase shift keying

As described for FSK, so four phase levels will provide two data bits per symbol in a phase shift keying scheme. Thus four-phase *quadrature* PSK (QPSK) provides twice the data throughput in the same bandwidth when compared to BPSK. On the other hand QPSK needs two 90° offset multipliers, as shown in Figure 6.12, each acting as a binary PSK modulator, as in Figure 6.6.

As shown in the diagram of QPSK generation, each symbol is formed by two data bits, which causes the resultant quadrature carrier signal addition to come to rest at 45° intervals around the phase diagram, shown as an insert in Figure 6.12. This combined vector state diagram shows where the signal phase is at the centre of each symbol period. The bandwidth required for transmission of QPSK can in theory be $0.5f_b$, but to allow for filtering as described above, and below, the bandwidth is increased to $0.7f_b$ in practice. This represents a modulation efficiency of 1.4 bps/Hz as compared to an ideal modulation efficiency of 2 bps/Hz.

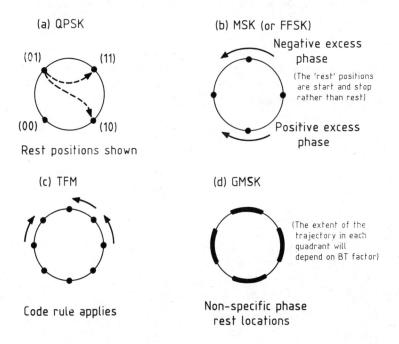

Figure 6.13   The movement of the carrier phasor during digital modulation signalling for the four modulations, QPSK, MSK, TFM and GMSK

Looking at the phase diagram of QPSK, the interesting question is what happens as one moves from one symbol (say 01) to the next (say 11). If there is no coherence between the data rate and the carrier frequency the waveform may well be constrained to move across the phase circle, or rapidly round the circle, shown now in Figure 6.13(a). However, by combining off-set PSK in the lower path (Q) of the QPSK modulator, Figure 6.12, with shaping of the data pulses so that they follow a half-sine waveform profile, the modified QPSK waveform, now called *minimum shift keying (MSK)*, acquires a constant envelope and continuous phase waveform, in fact equivalent to FFSK, as discussed in section 6.4.2 above.

6.4.4 Minimum shift keying

Minimum shift keying achieves the objective of having the carrier move from one phase state to the next, around the phase circle, as shown in Figure 6.13(b) in $\pi/2$ increments. To do this the frequency must either be ahead of the nominal carrier frequency, or behind; that is, either f_s or f_m. For FFSK the shift of the carrier frequency is $\pm$ 300 Hz, that is, one quarter the bit rate of FFSK. The so-called frequency deviation f_s to f_m (600 Hz) is equal to half the bit rate (1200 bps), giving m = 0.5. However, MSK still suffers from relatively poor adjacent channel sidebands due to its FSK-type spectrum.

The way forward is to constrain this change of frequency by filtering to improve the situation. The diagram Figure 6.14 demonstrates the point. The sudden switch in frequency, from say f_m to f_s, which causes the phase to move by $\pi/2$ radians in a symbol period, is smoothed out and some improvement on the out-of-channel power spectral density results, which is illustrated in Figure 6.15.

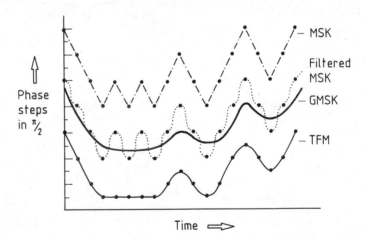

Figure 6.14   The phase movement with time as a function of the data for MSK, filtered MSK, TFM and GMSK

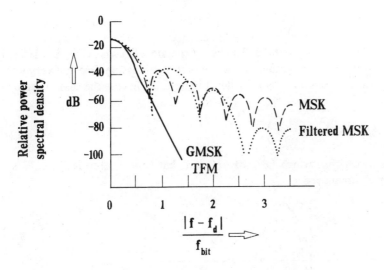

Figure 6.15   The adjacent channel power spectral density of MSK, filtered MSK, TFM and GMSK

For much greater improvement, but still retaining the important bandwidth efficient constant envelope property, two approaches have been developed.

6.4.5   Tamed frequency shift keying

The approach here is to code the data stream correlatively so that fewer and smaller phase changes take place. The code rule proposed established that a change of phase by $\pi/2$ only takes place if the three succeeding bits have the same polarity; whilst no phase change takes place if the three bits are of alternating polarity. Polarity changes of $\pi/4$ are reserved for the bit configurations 110, 100, 011 and 001. The result is a much more constrained movement around the phase circle, as depicted in Figure 6.13, and in Figure 6.14, with a correspondingly much improved spectral performance. This also accounts for the name, tamed FM (TFM). It must be recognised, however, that the phase rest positions are now placed at $\pi/4$ angles around the phase circuit, which must be taken into account in the receiver demodulator design.

6.4.6   Gaussian minimum shift keying

The classical theory of impulse transmission shows that transmitting at the Nyquist rate $f_b = B$, is *intersymbol interference* (ISI) free, because the ideal impulse response has regular zero crossings. Increasing the channel bandwidth to allow full cosine roll-off (twice bandwidth) retains the impulse shape, but also makes the far-off oscillations attenuate very rapidly. A compromise is to have a Gaussian-shaped filter, as depicted in Figure 6.16. This will generate a likewise Gaussian-shaped impulse response. There is some ISI at the first zero crossing time, but very little beyond. Thus a Gaussian-shaped filter, Figure 6.16, having the same envelope shape,

$$A(\omega) = \exp^{-0.54} \left(\frac{\omega}{\omega_c} \right) \qquad \qquad \dots (6.12)$$

produces a response which has only 1% ISI, but considerably better adjacent channel performance.

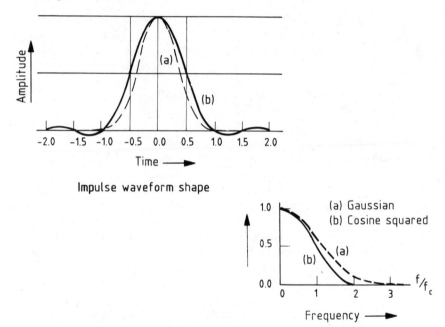

Figure 6.16   The impulse response of, (a) a Gaussian low-pass filter and (b) raised cosine-shaped LPF, shown as insert

Gaussian pulse pre-shaping can be added to the previous modulation MSK and then is termed *Gaussian* MSK (GMSK). The pulse shape after a Gaussian filter is also Gaussian, hence some ISI remains. One has a choice between the pre-modulation filter bandwidth B and the bit period T. If B > 1/T, then the waveform is essentially MSK; if, however, B < 1/T, the change of symbol cannot reach its next position in the time allotted. The effect is shown in Figure 6.13, when B < 1/T. The constellation becomes blurred due to the influence of remnants of previous pulses on the phase change. The phase behaviour attempts to follow the π/2 positions, as in Figure 6.14, but is constrained. In the receiver a complementary Gaussian filter will recover the data because it will again be influenced by the pre-designed intersymbol interference. It does mean however that the phase modulation pattern must be generated accurately, otherwise the phase pattern cannot be interpreted. A BT product, = 0.3, as B×T is denoted in the business, is planned for GSM (see Appendix, etc). GSM, as will be discussed below, operates with a bit rate of 270.833 kbps in a 200 kHz channel. This implies a modulation efficiency of 1.35 bps/Hz, that is, the bandwidth is just 0.74 times the bit channel rate, as expected for filtered QPSK.

It is also worth noting the so-called trellis diagram of these phase shift keyed modulations, shown in Figure 6.17. In the case of MSK, phase changes of π/2 occur at each data interval. All 1's cause the phase to move upwards at 45°; all 0's downwards at −45°. Other patterns will form the trellis.

In the case of GMSK, depending on the BT product value, the phase change is unable to follow the MSK trellis. This is why in the GSM signalling protocol, as described in Chapter 8, a burst data mode is necessary in order to maintain a phase reference, which would become uncertain after a long string of 1's or 0's.

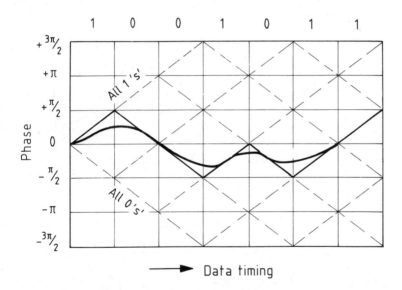

Figure 6.17   The trellis diagram of the carrier phase positions formed by data changes during MSK and GMSK modulation

Measurements confirm that the power spectral density of GMSK does indeed exceed that of MSK in respect of adjacent channel performance and a 60 dB out-of-band performance ratio can be achieved. GMSK (BT product = 0.3) is in theory a constant amplitude waveform; unfortunately, the fact that the carrier must be power controlled and also confined to timed bursts, implies that the overall carrier envelope will cause some spectral spreading when using class-C transmitter stages.

6.4.7   *Differential phase shift keying*

As we noted when discussing the generation of QPSK, Figure 6.12, a rectangular bit stream leaves one with a phase circle diagram with specific rest positions at the 45° corners, the actual carrier waveform exchanging phase at each symbol period. If the following coding scheme is used

Message symbol	00	01	10	11
Phase change	0°	-90°	+90°	180°

the signal space diagram shown in Figure 6.18(a) will be observed. The signal waveform must at intervals pass through zero, which is of course very difficult to amplify accurately, especially within a transmitter.

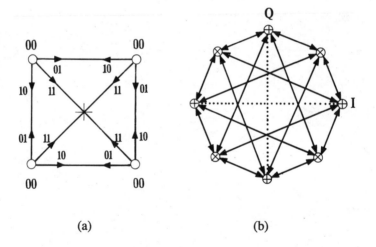

(a)   (b)

Figure 6.18   Signal space diagrams for QPSK, (a) conventional, (b) $\pi/4$ differential

Also it has the disadvantage that for the symbol 00 there is no phase change and the signal vector will rest at one of the nodes.

If differential encoding is applied to the data stream, such rest positions do not occur because the transmitted data is differential. Differential encoding applied to binary PSK is shown in Figure 6.19. If this scheme is now applied to QPSK generation, as in Figure 6.20, and made four-level, as shown, a totally new signal space diagram is formed, which is constructed in Figure 6.18(b).

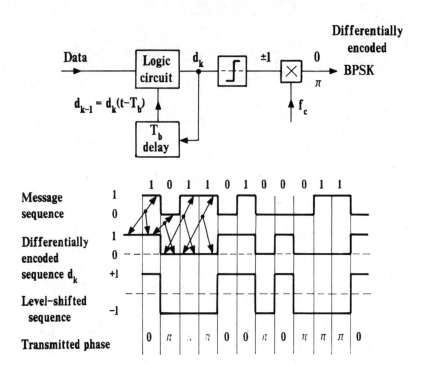

Figure 6.19   Differential encoding applied to a BPSK modulator

This uses the coding rule:

Message symbol	00	01	10	11
Phase change	−45°	−135°	+45°	+135°

and the pattern shown in Figure 6.18(b) comes about. The carrier vector now has no rest position; nor does it pass through zero. Some amplitude variation does occur and quasi-linear amplification is necessary in a DPSK transmitter.

This is the modulation to be adopted for the North American digital cellular plan. Here a raised cosine-shaped filtered $\pi/4$ shifted DPSK system will be used. A transmission rate of 48.6 kbps in a channel spacing of 30 kHz is planned, which gives a spectrum efficiency of 1.62 bps/Hz, a 20% improvement over GSM, at least in this context.

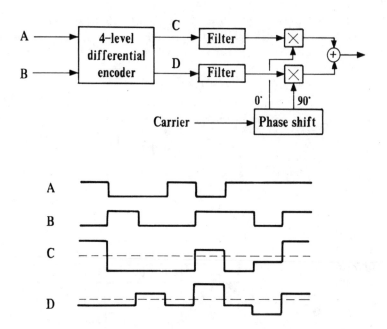

Figure 6.20   Four-level differential encoding applied to a QPSK modulator so as to generate DPSK

As may be well known, it is practical to generate a multi-level PSK waveform, either having the phase reference positions placed symmetrically around a circle in the signal space diagram (M-ary PSK), or quaternary amplitude shift keying waveform (QAM).

Because, as was discussed earlier, multipath propagation becomes much more disruptive on multi-level modulations, they have so far not been usually considered suitable for cellular schemes; hence M-ary modulations, where M > 4, will not be discussed.

6.5   Bit error rate

Digital modulation does not suffer an apparent worsening S/N ratio, as does analog modulation since, as explained in Chapter 1, digits become misinterpreted and an increasing bit error rate (BER) response sets in. The probability of an error being recorded, when judged against the received carrier-to-noise ratio, can be calculated by considering the error region surrounding the signal vector. Thus Figure 6.21 shows the signal vector space for QPSK and in particular the signal vector (11).

An interfering signal $V_i \geq V_c / \sqrt{2}$ will clearly cause the symbol to be misinterpreted. One can compute (with some difficulty) the probability of making an error versus the carrier-to-noise ratio C/N, but it is more efficient to express C/N as the ratio of the E_b/N_b as follows:

Let carrier power C = energy per bit $E_b \times$ rate R, $(= f_b)$

noise power N = noise power per Hertz $N_o \times$ B, $(= B_w)$

$$\therefore \qquad \frac{C}{N} = \frac{E_b}{N_o} \times \frac{R}{B}$$

$$\text{or } E_b N_o \qquad \frac{E_b}{N_o} = \frac{C}{N} \times \frac{B}{R} \qquad \qquad \dots (6.13)$$

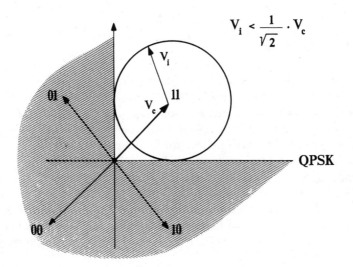

Figure 6.21   A QPSK signal vector space in the presence of an interfering vector

The bit error probability P_e then has the general form:

$$P_e = \frac{1}{2} \, \text{erfc} \left(\frac{E_b}{N_o} \right)^{\frac{1}{2}} \qquad \qquad \dots (6.14)$$

The shape of P_e versus E_b/N_o is also shown in Figure 6.22.

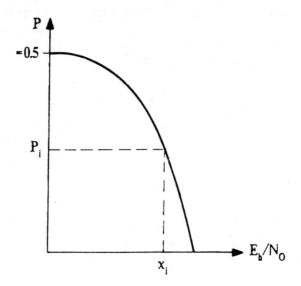

Figure 6.22   General shape of digital transmission BER in a white noise
environment

A specific comparison of BPSK and QPSK (which is 4-level PSK) is interesting, because as shown in Figure 6.23, QPSK is some 3 dB worse in performance
than BPSK. However, the bit rate is twice in the case of QPSK, so that if only
additive white Gaussian noise (AWGN) was the cause of errors, multi-level
modulations would perform as well as two-level binary modulation.

In mobile radio several other sources of channel imperfection arise, however.
Thus whereas in Figure 6.24 the theoretical result is shown, in practice, receiver
and transmitter RF circuits are added to the data transmission system, and also
multipath will exist in the propagation environment. The observed BER worsens,
as shown, and is often limited to some *irreducible* level. These three cases are
marked laboratory, static and moving at 30 mph, in Figure 6.24, respectively;
see also Figure 1.13.

These imperfections can easily amount to 6dB or more. Carrier-to-noise ratio
therefore needs to be this much higher to achieve the same BER.

When digital modulation schemes form part of a mobile system, multipath
propagation causes further degradation in the system performance and necessitates
channel sounding and path equalisation techniques, which will be referred to again
in Chapter 8.

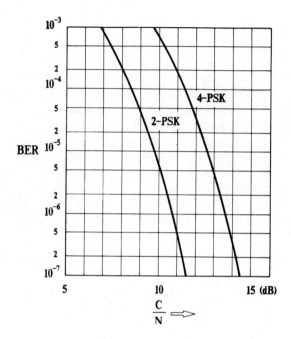

Figure 6.23   BER versus C/N for ideal PSK in an AWGN environment

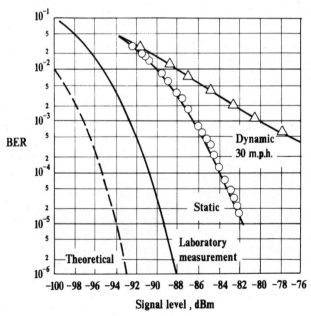

Figure 6.24   Expected and observed bit error rate performance for 2400 bps
                MSK over a 12.5 kHz VHF radio channel at 172 MHz

6.5.1 Improving BER

Another important consideration in regard to digital modulation is the form of demodulation. Coherent methods (which require recovery of the precise carrier frequency and possibly phase) are generally superior to the more easily implemented non-coherent circuit arrangement, such as an FM discriminator. Figure 6.25 illustrates the point and, for example, DPSK can offer approximately a 6 dB improvement at the more critical BER levels, such as BER $= 1 \times 10^{-3}$.

From a digital cellular radio design point of view, BER versus C/N ratio is clearly more relevant, more so than the received signal level in dBm. Recalling (6.13), if C is expressed in dBm and using $N = -174$ dBm $+ 10 \log B$, plus receiver noise figure F in dB, the ratio B/R will depend on the spectral efficiency of the modulation and, as we observed, a high efficiency modulation will usually make the ratio E_b/N_o worse.

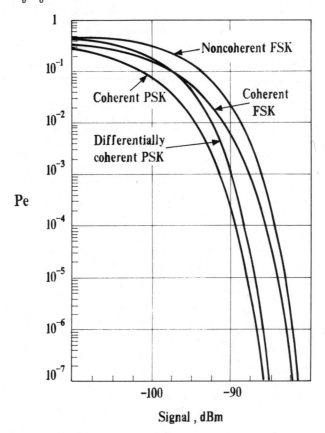

Figure 6.25   BER performance of four binary modulations versus relative signal level

Data errors can also be reduced, or corrected, by forward error correction (FEC) techniques. Unfortunately, FEC implies transmitting more data, which means, either (i) accepting more errors, or (ii) lowering the data transmission rate, or (iii) increasing the signal power. The last implication is contained in Shannon's formula

$$\frac{R}{B} < \log_{10} \left[1 + \frac{E_b R}{N_o B} \right] \qquad \ldots (6.15)$$

This says that if the bandwidth is limited, one can only increase the transmission rate provided that the minimum E_b/N_o ratio increases faster.

The way around the conflict is to combine the coding and modulation. The data message is now pre-coded before modulation in specific patterns, such as a set-partitioned code, and this code can be recognised with added confidence in the received signal. Hence, use of the code saves transmitter power on a per-data bit basis without increasing the signal bandwidth. *Coding gains* of up to 6 dB in E_b/N_o ratio have been predicted. So-called *trellis coding* leading to so-called *trellis code modulations* (TCM) are mainly used, based on the phase locations which are specific to MSK type of modulation indicated in Figure 6.17. Such strategies form part of the GSM system implementation described in Chapter 8.

Further reading

Bennett, W.R. and Davey, J.R. (1965). *Data Transmission,* McGraw-Hill Book Co, NY

Couch, L.E. (1987). *Digital and Analog Communication Systems*, Macmillan, NY

de Jager, F. and Dekker, C.B. (1978). 'Tamed frequency modulation, a novel method to achieve spectrum economy in digital transmission', *IEEE Trans.,* COM-26, May, p. 534

Feher, K. (Ed) (1987). *Advanced Digital Communications,* Prentice-Hall, NY

IEEE J. Selected Area Comms, (1989). 'Bandwidth and power efficient coded modulation', Aug. and Dec. issues

Murota, K. and Hirade, K. (1981). 'GMSK modulation for digital mobile radio telephones', *IEEE Trans.,* COM-29, No. 7, July, p. 1044

Sklar, B. (1988). *Digital Communications; Fundamentals and Applications,* Prentice-Hall International, Inc, NY

Sunde, E.D. (1954). 'Theoretical fundamentals of pulse transmission', *BSTJ*, 33, May, p. 721 and Pt II, July, p. 987

Ziemer, R.E. and Tranter, W.H. (1990). *Principles of Communication Systems, Modulation and Noise,* Houghton Mifflin, Boston, USA

7 Speech Coding

7.1 Introduction

It is clearly necessary to find a way of digitizing speech if an all-digital cellular radio system is to be achieved.

Pulse Code Modulation (PCM) is now used throughout the fixed telephone network, being first introduced around 1962. Many studies have dealt with the standard 64 kbps PCM voice transmission hierarchy, which assumes 8 bits per sample at a sampling rate of 8000 samples per second. Using non-linear sampling improves the inherent signal-to-noise ratio so that a 64 kbps log-law PCM codec is a CCITT standard, known as G 711.

Speech coding has come a long way since the intoduction of pulse code modulation. Present-day techniques seek to exploit the intrinsic properties of speech signals in order to remove redundancy and achieve good speech quality at much lower bit rates. This chapter is basically a brief introduction to speech coding, with emphasis being given to the encoding of narrowband telephone signals for network communication quality speech. The important coding techniques and achievements are discussed in the light of existing international coding standards.

7.2 Coding requirements

Speech coding algorithms are developed and optimised to satisfy a number of application specific requirements. Obviously, the quality of the recovered speech signal plays a major role in the design of a codec. The objective signal-to-noise ratio (SNR) can assist in evaluating and comparing the performance of systems operating at relatively high bit rates (> 16 kbps). However, when SNR measures are applied to intermediate and low bit rate codecs they often fail to correlate well with the subjective quality of the decoded speech signal. As a result various subjective tests are employed to quantify output speech quality, with the *mean opinion score* (MOS), scale (1 to 5), test being the most popular.

When applied to narrowband telephone speech (300 Hz to 3400 Hz) an MOS score greater than or equal to 4.0 implies network quality speech (often referred to as 'toll' quality). An MOS value in the range of 3.5 to 4.0 corresponds to communication quality, and is characterised by some degradation, noticeable by a listener. Communication quality speech is acceptable in certain telephone applications, such as mobile radio and voice mail. MOS values in the region of 2.5 to 3.5 imply synthetic quality speech. This is highly intelligible, but with reduced naturalness and limited speaker recognisability. Synthetic quality is found in low bit rate (2.4 kbps) secure voice transmission systems.

Decoded speech quality is closely linked to two other codec design factors, namely output bit rate and codec complexity. In general, as the required bit rate decreases, codecs become more sophisticated and consequently more complex, in an attempt to retain high decoded speech quality. Obviously, algorithm complexity has a direct impact on implementation cost and power consumption. Nevertheless, speech coding quality deteriorates at lower bit rates and the challenge is to design efficient low bit rate, high speech quality, low complexity codecs. Figure 7.1 highlights speech quality as a function of output bit rate for the three classes of speech coding schemes, namely waveform coders, vocoders and hybrid systems.

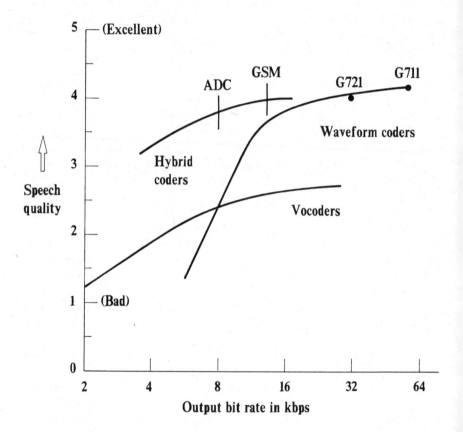

Figure 7.1   Speech quality versus bit rate for telephone bandwidth speech codecs

7.3 Coding techniques

7.3.1 Waveform coders

Speech coding algorithms have been developed for encoding what is termed telephone bandwidth speech, and can be classified into three categories. The first class consists of algorithms which attempt to reproduce, at the output of the decoder, a close approximation of the original speech signal. They are known as waveform coders and operate successfully at intermediate and high bit rates (2 to 8 bits/sample). These coders are now briefly described.

Pulse code modulation: as already mentioned, the first waveform coding technique to be developed consists of both time quantization (sampling) and amplitude quantization and, although of limited compression/quality capability, PCM is used today at 64 kbps, in both public and private 'fixed' telecommunication networks, in the form of the G711 CCITT standard, with an MOS score of 4.3. The amplitude quantizer (or simply the quantizer) is an important element present in all encoding systems.

Differential coders form an error signal, as the difference between successive input speech samples and a corresponding prediction estimate, which is then quantized and transmitted. *Adaptive differential pulse code modulation* (ADPCM) and *adaptive predictive coding* (APC) represent two important intermediate bit rate (32 to 16 kbps) differential codecs. Both systems estimate the incoming input samples using previously decoded samples.

ADPCM employs a short-term predictor which models (partially) the speech spectral envelope. The predictor can be forward or backward adaptive, with its coefficients defined periodically (block adaptive) or at every sampling instant (sequentially adaptive).

The CCITT G721 ADPCM standard, introduced in 1984, achieves network quality speech (MOS score of 4.1) at 32 kbps. This is a low complexity codec of reasonable robustness, when operating with channel bit error rates in the range of 10^{-3} to 10^{-2}, and is therefore well suited for wireless access applications based on low power, handheld cordless telephones.

This standard has been extended in recommendation G723 to operate at 24 and 40 kbps. In addition, an *embedded* ADPCM standard has been established by CCITT, which operates at 40, 32, 24 and 16 kbps and can be used in wideband packet network applications. Notice that because speech quality deteriorates considerably at bit rates below 32 kbps, noise shaping and post-filtering needs to be added to the codec in order to minimise the perceptual effect of quantization noise.

APC employs both short and long-term prediction in a differential coding structure. The system outperforms ADPCM at 16 kbps and offers communication quality speech at bit rates as low as 10 kbps. As with ADPCM, the introduction of noise shaping and post filtering in APC reduces the subjective loudness of quantization noise. ADPCM is proposed for the digital European cordless telephone (DECT) office system, but it does have the penalty of complexity of implementation and power consumption.

Delta modulation, that is, a one-bit PCM variant, in the form of *digitally variable slope delta modulation* (DVSDM) is used in one version of CT2 (a second generation UK cordless telephone) because of its lower simplicity and cost, though of lower quality.

7.3.2 Vocoders

The second class of speech coding techniques consists of algorithms called *vocoders* which attempt to describe the speech production mechanism in terms of a few independent parameters serving as the information-bearing signals.

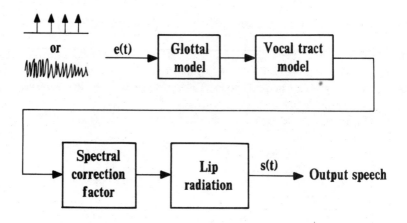

Figure 7.2   The source/filter model for speech

Vocoders consider that speech is produced from a 'source-filter' arrangement, such as set out in Figure 7.2. Voiced speech is the result of exciting the vocal tract (filter) with a series of quasi-periodic glottal pulses generated by the vocal cords (source). Unvoiced speech, on the other hand, is produced by exciting the filter with random white noise. Thus, vocoders operate on the input signal, using an 'analysis' process based on a particular speech production model, and extract a set of source-filter parameters which are encoded and transmitted. At the receiver, they are decoded and used to control a speech synthesizer which corresponds to the model used in the analysis process. Provided that all the perceptually significant parameters are extracted, the synthesized signal, as perceived by the human ear, resembles the original speech signal.

The 'filter' part of the speech production model can be defined by operating in the time or frequency domain and effectively determines the 'envelope' information of the speech spectrum.

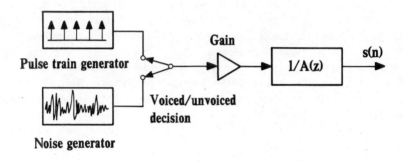

Figure 7.3   The decoding arrangement of the basic vocoder

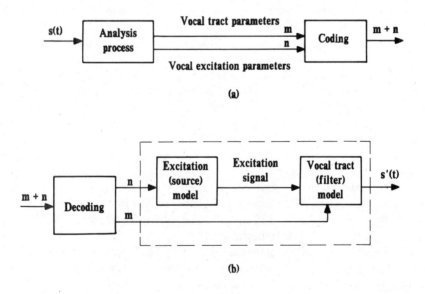

Figure 7.4   Generalised block diagram of a vocoder, (a)   analysis at the transmitter, (b)   synthesis process at the receiver

The most popular model today is the so-called *linear predictive coder* (LPC) and can be viewed as a multi-stage filter A(z) set up by a digital (time domain) algorithm. Figure 7.3 indicates how the vocal model of Figure 7.2 has been reduced to a single-filter function with either a pitch impulse generator for voiced frames, or a noise generator for unvoiced frames. The source power is set by the amplifier. This circuit represents the decoding side of the vocoder.

The digitized speech waveform now no longer needs to be transmitted from the speaker to the listener; all that is required are the parameters of the filter, the gain of the amplifier and when to use the voiced or unvoiced source. A generalized block diagram of a vocoder is shown in Figure 7.4.

Vocoders are medium complexity systems and operate at low bit rates, typically 2.4 kbps, with synthetic quality speech. Their poor quality speech is due to, (i) the óversimplified 'source' model used to drive the 'filter' and, (ii) the assumption that the source and the filter are linearly independent.

7.3.3   Hybrid coders

The poor and synthetic quality of speech vocoders has led to what is known as the *residual excitation* approach to speech coding, a concept whereby not all the speech is synthesized, but a small part is transmitted as a coded waveform part of the original envelope, hence the name hybrid. The penalty is the higher bit rate of transmission required, but a very much improved speech quality is realized.

The system extracts a low frequency band from the input signal (typically 300 to 940 Hz) which is waveform coded and transmitted, in addition to the vocoder channel (filter information). This baseband signal, which contains the required excitation information, is processed at the receiving end by a non-linear element that flattens and broadens the signals spectrum, without affecting its periodicity (if any), to yield an improved excitation signal. However, the improved speech quality obtained from the voice excited channel vocoder (and from other residual excitation systems) is achieved only at the expense of several extra kbps needed to code the baseband (residual) signal.

Many residual excited hybrid coding systems have been proposed, most of them using linear predictive modelling of the synthesis filter. In particular, a *residual excited linear predictive* (RELP) coding system has been developed for low to intermediate bit rate (4.8 to 16 kbps) operation.

RELP systems employ short-term (and in certain cases, long-term) linear prediction, to formulate a difference signal (residual) in a feed-forward manner. Early systems used baseband coding and transmitted a lowpass version of the residual. The decoder recovered an approximation of the full-band residual signal, by employing high frequency regeneration which was subsequently used to synthesize output speech.

Multipulse-excitation linear predictive coder (MPE-LPC) systems model the excitation signal as a sequence of irregularly spaced pulses. MPE coders employ a synthesis filter which consists of one or two autoregressive (AR) filters in series. The first filter models the 'smooth' spectral envelope of the signal (short-term filter) while the second (if used) models the harmonic (fine) structure of the spectrum (long-term filter). The parameters of the excitation model (i.e. the pulse positions and amplitudes) and part of the synthesis filter (i.e. the long-term filter) are determined in a closed-loop optimization process.

An outline of an MPE decoder is shown in Figure 7.5. The multipulse excitation generator can be seen. Also shown is a block marked *long-term predictor* (LTP). Because the speech to be coded is divided up into segments (usually 20 msec blocks) the coder would have no long term memory to assist with the excitation (often pitch), nor as just discussed, a predicted model for the filter. Thus the design of the LTP is critical to these hybrid coders.

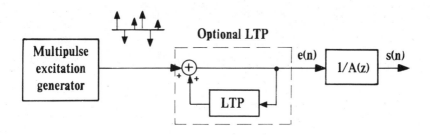

Figure 7.5   Outline of a multipulse excited linear predictive decoder

MPE-LPC coders provide near network quality speech (MOS = 4) at bit rates in the range of 16 to 8 kbps. Their performance deteriorates rapidly, however, at bit rates below 8 kbps where acceptable performance can only be achieved by drastically modifying the basic multipulse excitation model.

A special case of MPE-LPC coding is the *regular-pulse excitation* (RPE) LPC coder which models the excitation signal with a sequence of equally spaced pulses. The performance of RPE systems is similar to that obtained from MPE coders. Various computationally efficient RPE schemes have been proposed and one of them has been chosen as the coding standard for the 'full-rate channel' of the GSM European mobile radio system. The codec operates at 13.2 kbps with a delay of 40 msec, a reasonably robust performance in the presence of channel errors, and can be implemented on a single DSP device.

The GSM speech coder is illustrated in its basic form in Figure 7.6. With an output rate of 13 kbps, the encoder processes 20 ms blocks of speech, and represents each block with 260 b. It is helpful to observe three functional parts of the encoder, which perform the linear predictive analysis, long-term prediction, and excitation analysis, respectively. The linear predictor is an eight-tap filter characterised by eight log-area ratios, each represented by 3, 4, 5, or 6 b. In aggregate, the eight log-area ratios are represented by 36 b. Likewise, within the 20-ms block, the long-term predictor estimates pitch and gain four times (at 5 ms intervals). Each estimate produces a lag coefficient (7 b) and a gain coefficient (2 b). In total, the long-term predictor generates 36 b over the 20 ms block. The remaining 188 b come from the regular pulse excitation analysis which, like the long-term predictor, operates over 5 ms sub-blocks, producing 47 b per sub-block.

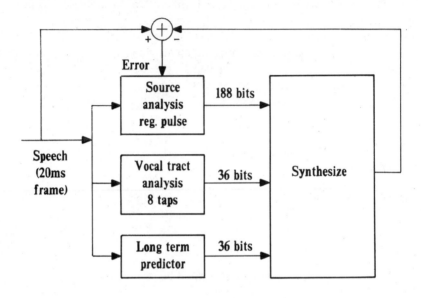

Figure 7.6   Regular-pulse excitation speech coder

For purposes of error control, the speech coder has two categories of output bits. The 182 'class 1' bits have a stronger influence on speech quality and are protected by three error-detecting parity bits and a half-rate convolutional code, producing a sequence of 378 channel bits. As shown in Figure 7.7, this sequence is combined with the 78 unprotected class 2 bits to form the 456 b speech signal. The aggregate bit rate per speech signal, which was 182 + 78 (= 260 b), is 456 b in 20 ms, = 22.8 kbps.

In a GSM receiver the 260 bits have to be unravelled to operate the appropriate functions. The details are of course laid down in the full GSM specification referred to in the next chapter, but are too specific for full discussion here. Figure 7.8 outlines the GSM RELP decoder and indicates the specific radio signal bit data allocation. Working from left to right, three pulse groups set up the appropriate RPE source, being synthesized directly from the low frequency tones of the original speech. Moving across, a further group of pulses control the LTP module. Finally, the bulk of the data is used to set up the parameters of the LPC synthesis filter, organised in a manner so as to require the least number of pulses (per speech segment). Pre-emphasis and de-emphasis are used as for FM analog modulation systems.

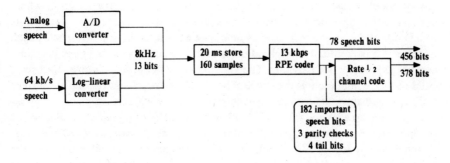

Figure 7.7   GSM source and channel coding

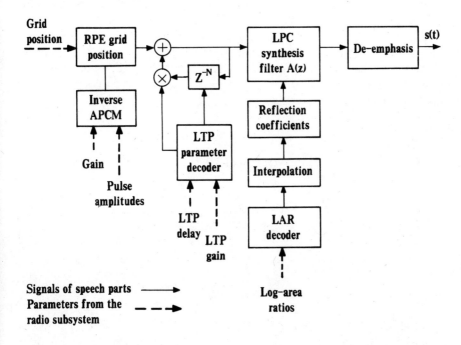

Figure 7.8   Outline of GSM-RELP digitized speech hardware decoding strategy

7.3.4   Codebook vocoders

Codebook-excitation linear predictive (CELP) coders employ a vocal tract LP-based model, a codebook based excitation model and an error criterion which serves to select an appropriate excitation sequence using an analysis-by-synthesis (AbS) optimization process. The system selects that excitation sequence which minimises a perceptually weighted mean square error formed between the input and the locally decoded signals. The vocal tract model utilises both a short-term filter (STF), which models the spectral envelope of speech, and a long-term filter (LTF), which accounts for pitch periodicity in voiced speech.

The main (fixed) excitation codebook was originally designed as a collection of random vectors (sequences), each of which is constructed using samples from a set of independent identically distributed Gaussian random variables having zero mean and unit variance. In this case the number of computations required to select the 'optimum' excitation sequence is prohibitively large and thus various simplified search strategies suitable for real-time implementation have been proposed. These codebook simplifications are based on efficient random codebooks, with a modelling performance that is equivalent to that of the Gaussian random codebook, but using structured and multistage codebook search strategies.

An outline of a CELP decoder is shown in Figure 7.9. Note the similarly to Figure 7.5, except that now the excitation is chosen from preprinted (codebook) models. More naturalness over and above RELP, or MPE-LPC, is possible for the same bit rate, or perhaps more importantly, a lower bit rate coder is acceptable for operational performance.

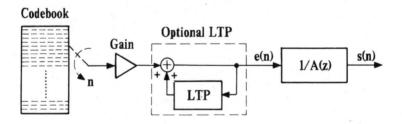

Figure 7.9   Outline of the CELP decoder

In addition to the above modifications, whose aim is to improve computational efficiency, CELP speech quality can be enhanced by employing long-term filters with high temporal resolution and/or some form of post-filtering. CELP coders provide good toll quality speech at 8 kbps, with a typical MOS score of 3.7, while at 4.8 kbps CELP is far more successful than MPE-LPC in producing communication quality speech. The *vector sum excited linear predictive* (VSELP), CELP system has recently been confirmed as the 8 kbps North American standard for cellular telephony. The same algorithm has also been adopted at 6.7 kbps for the Japanese digital mobile radio system. The USA DoD (Department of Defense) 4.8 kbps speech coding standard is also a CELP-type system with speech quality comparable to 32 kbps CVSD speech.

The block diagram of the VSELP speech coder has the same form as Figure 7.6, i.e, for GSM, but the details of the excitation analysis and bit assignments differ. With a source rate of 7.95 kbps, the North American code is more efficient than the 13 kbps code of GSM. On the other hand, the signal processing hardware of the coder is more complex.

The coder derives a new linear prediction every 20 ms. This linear predictor is characterised by ten log-area ratios, which are represented in aggregate by 38 b. The frame energy accounts for 5 b. The coder obtains long-term predictor coefficients at 5 ms intervals, four times per 20 ms block. The lag of each long-term predictor is represented by 7 b, accounting for 28 b per 20 ms block. As a form of codebook-excited linear prediction, the VSELP coder computes 14 b of excitation information in every 5 ms sub block. It also computes 8 b of gain information, which together represent a scaling factor for the long-term predictor and two scaling factors for the vector sum codebook. All this adds up to 159 b per 20 ms, or 7.95 kbps.

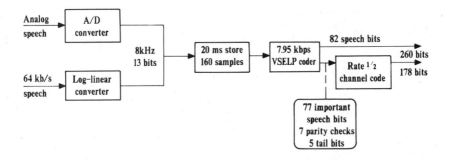

Figure 7.10   VSELP source and channel coding

As in GSM, the output bits of the speech coder are divided into two categories, of which the 77 class 1 bits have a greater influence on speech quality than the 82 class 2 bits. The class 1 bits are protected by an error-detecting code and a half-rate convolutional code to produce a sequence of 178 channel bits, which are multiplexed with the 82 class 2 source bits to produce the 160 transmitted bits, as indicated in Figure 7.10. Two blocks of 260 channel bits are interleaved and placed in the two assigned time slots of the 40 ms transmission frame.

Further reading

Atal, B.S. (1982). 'Predictive coding of speech at low bit rates', *IEEE Trans Communications,* COM - 30, April, pp 600-614

Bellamy, J. (1990). *Digital Telephony,* Wiley, USA

Flanagan, J.L. (1972). *Speech Analysis Synthesis and Perception,* Springer-Verlag, Heidelberg

Goodman, D.J. (1991). 'Second generation wireless information networks', *IEEE Trans Vech Tech,* VT-40, May, pp 366-374

Jayant, N.S and Noll, P. (1984). *Digital Coding of Waveforms,* Prentice-Hall, USA

Natvig, J.E. (1988). 'Evaluation of six medium bit-rate coders for the pan-European digital mobile radio system', *IEEE Journ. Select Area Comms,* 6, Feb, pp 324-331

Singhal, S., LeGall, D. and Chen, C.T. (1990). 'Source coding of speech and video signals', *Proc IEEE,* 78, July, pp 1233-1250

Xydeas, C. (1991). Speech coding, in *Personal and Mobile Radio Systems,* Peter Peregrinus Ltd, UK

8 Digital Cellular Designs

8.1 Second generation networks

Analog cellular is known as a first generation wireless information telephone network. There are up to six incompatible analog cellular standards worldwide (summarized in the Appendix), even though they have the underlying FM FDMA mode of operation. The main aim of the so-called second generation systems was (i) to go digital, (ii) to have a single standard. The first is taking place; but the second is unlikely to be achieved. A common mode of operation will be the feature of *time division multiple access* (TDMA), but matters like frame timing, method of modulation and error-correction code procedures will differ from one global system to another. The three main systems at the present time are:

- GSM - Global system for mobile communications
- JDC - Japanese digital cellular
- ADC - American digital cellular

These second generation systems have reached a highly advanced state of design and trialling. The principles can therefore be described in some detail and with reasonable accuracy. The details of GSM are, however, the most readily available and so emphasis will be put on that system. Again, it is the availability of the radio spectrum (by allocation) which has dictated the way in which the TDMA frame arrangements have come about.

Third generation systems are also being researched, using methods such as *code division multiple access* (CDMA) and controlled packet allocation. These considerations at present are well outside the scope of this text and any agreed cellular radio system designs.

Cellular radio as a network does not need to specify the way individual subscribers are accessed by the network. For example, first generation systems use the familiar one-to-one mode of conversation, the single-channel-per-user concept and frequency division multiple access.

We recall that Figure 3.6 shows the FDMA arrangement in association with trunking and the use of a control channel.

In this case the available spectrum is divided into channels A, B, C, D, ... The bandwidth of each channel is set by the transmitter emission mask, namely Figure 8.1. It is seen that the bulk of the transmitted (or received) power occupies less bandwidth than the allocated bandwidth. In the case of the analog TACS system, although the allocated channel bandwidth is 25 kHz, the FM modulation indexes used for voice and signalling widen the signal bandwidth and the signals spill over into the adjacent channel. This conflict between allocated bandwidth and occupied bandwidth is one of the arguments put forward to claim that FDMA is not efficient in the use of the spectrum; see the earlier discussion in section 6.2.

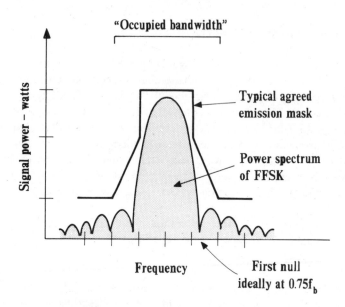

Figure 8.1   Transmitter emission characteristics

With FDMA a single user will occupy each channel, and this is true whether the modulation is analog or digital.

The signalling is already digital, being FFSK in most cases. Therefore, it would be quite straightforward to introduce digital speech so that each channel becomes a continuous digital stream when in use. If the modulation method used has an efficiency of 1 bps/Hz, then in principle fully protected RELP encoded speech with a transmission rate of 22.8 kbps could be carried by the 25 kHz channels (ignoring any signalling). One would then have a digital cellular radio scheme using FDMA.

The aim of going digital is to give compatibility with the fixed network supporting the cellular radio system; nevertheless, second generation networks have proceeded, based on an alternative access method.

8.2   Time division multiple access

An alternative way of using the available (allocated) spectrum is to let each user have access to the whole band for a short time (traffic burst), during which time he transmits data much faster. He shares his frequency allocation with the other users who have time slot allocation at other times. Figure 8.2 shows a TDMA arrangement where the spectrum is only *partially* allocated, each user group A, B, C, D, ... having a time slot allocation in a particular channel group, 1, 2, 3, ...

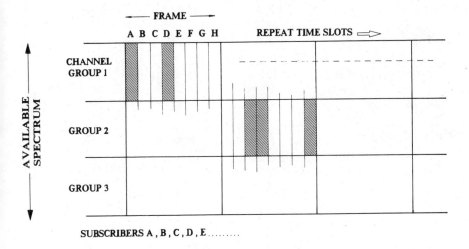

Figure 8.2   The TDMA mode of operation: each channel group of eight subscribers (A....H), send (and receive) messages as bursts. The number of base station transmitters is equal to the number of groups

This arrangement is sometimes known as *narrowband* TDMA. If all of the available spectrum (say, the forward channel - a similar picture will apply to the reverse channel) was allocated to each user during his time slot, this situation would be *wideband* TDMA. Each user would have to transmit data very rapidly, i.e. a high data rate, in order to pass traffic. Alternatively, more leisurely frames could be set up within the same channels, as apply to an existing FDMA scheme. The ADC and JDC systems intend to begin with just three time slots per frame. The relationship between the number of subscribers per group and frame length, etc., can be ascertained from Table 8.1.

Variation within the tabular data also occurs due to the speech coding rate assumed and the modulation efficiency. These two matters were discussed in the two proceeding chapters.

For a chosen bit rate one could, as an exercise, calculate the various number of frames and frame length periods for various TDMA strategies. In practice, however, the systems being implemented are so tightly specified, plus the fact that many other factors in relation to the data within a frame have to be taken into consideration, that the exercise would be redundant.

Table 8.1   FDMA to TDMA changes

SYSTEM	TACS	ADC	GSM
Multiple access	**FDMA**	**TDMA**	**TDMA**
Channels per carrier	1	3	8
Carrier spacing (kHz)	25	30	200
No. of carriers	400	333	50
No. of channels	400	999	400
Frame period (msec)	no limit	40	4.6
Channel data rate (kbps)	10	48	270

For the actual frequencies used for the forward and reverse channels see the Appendix Table A.1. The table here assumes 10 MHz of spectrum available for each service in a frequency division duplex mode.

The breakdown of frames into time slots with the individual message content is shown in Figure 8.3. Several variations of this arrangement are possible.

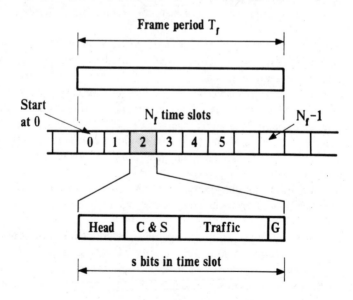

Figure 8.3   The frame, time slot and message relationship in TDMA

Here N_f channels per frame are shown. The convention is to start with number 0. Each time slot, belonging to an individual user, is made up of several parts, namely

Header message:	This contains carrier recovery; bit timing recovery; unique words; channel identity
C and S:	Control and signalling
Traffic:	This is the message part to the subscriber; typically, it would be encoded
G:	Guard space, to allow for time/distance delay, because of cell size

From Figure 8.3 it can be seen that much of the activity of a TDMA system is taken up with what is called identification and authentication. The overall bit rate allocated to each user is only a percentage of that available in a single-person-per-carrier scheme; on the other hand, no channels are uniquely allocated as control channels.

In TACS the control channels can interpreted as reducing the spectrum efficiency by 21/300 or 7%. Also, there is the difficulty of not being able to allocate alternate channels. In the TDMA schemes to be described, the efficiency will depend on the number of non-traffic bits.

8.2.1 Possible advantages of TDMA

What then are the advantages of TDMA to a cellular radio scheme?

- *Multiple circuits per carrier.* As indicated, all TDMA formats seek to multiplex at least two, and more often four, or eight, circuits per carrier. This can represent lower investment at each base station site because of saving on transmitters (see Figure 3.5).
- *Burst transmission.* The transmission from the mobile units is not continuous, but occurs during specified time slots only. This has many implications for circuit design and system control. It also impacts, positively, on the co-channel interference equation (3.2), since at any given moment only a percentage of the mobile units in operation are actually transmitting, leading to better frequency reuse (see Figure 3.13).
- *Bandwidth.* The bandwidths of proposed TDMA systems range from 20-30 kHz, equivalent to analog channels, to more than ten times this value. Bandwidth is determined in part by the choice of modulation technique; but to increase the efficiency, multi-level modulations would have to be used, or half-rate coders. The higher channel rates proposed for some TDMA systems will create a much more severe intersymbol-interference problem than FDMA systems. For a 270 kbps rate, for example, the symbol time is 3 μs, which is

approximately the same as the delay spread that is expected in dense urban centres. Even using four-level modulation to attain a spectral efficiency of 2 bps/Hz would only increase the symbol time to 6 μs. Adaptive equalisation is necessary. Some TDMA units may also be required to do slow-frequency hopping, to improve resistance to multipath fading. For TDMA systems with lower channel rates, however, the equalisation requirement may be no more severe than for FDMA systems.

• *Higher mobile-unit complexity.* The TDMA mobile unit has more to do than the FDMA unit, at least on the digital-processing side. Whether this is a crucial difference depends upon the processing-device technology that is assumed. With the continuing advances in very large scale integrated circuitry, the added processing needed for TDMA grows less significant.

• *Higher transmission overhead.* A TDMA slotted transmission forces the receiver to reacquire synchronization on each burst. Also, guard periods may be necessary to separate one slot from another, in case unequalised propagation delay causes a distant user to slip into the adjacent slot of a nearby user. Because of this, TDMA systems usually need much more overhead than FDMA systems. The conventional view is that TDMA overhead requirements can run to 20-30% of the total bits transmitted, which is indeed a significant penalty. On the other hand, the measurement of path delay from frame to frame does imply that more efficient handover procedures are possible.

• *Lower shared-system costs.* As already indicated, the major advantage of TDMA systems over FDMA arises from the fact that each radio channel is effectively shared by a larger number of subscribers. The cost of the central-site equipment is significantly reduced.

• *No duplexers required.* An important advantage for TDMA over FDMA arises from the fact that by transmitting and receiving on different slots it is possible to eliminate the duplexer circuitry in the MS, replacing it with a fast-switching circuit to turn the transmitter and receiver on and off at the appropriate times.

• *Openness to technological change.* TDMA systems have another feature over FDMA which may outweigh everything else, at least to system designers and/ or regulators who look to the long-term technological viability and flexibility of the system. As bit rates fall for speech coding algorithms, a TDMA channel is more easily reconfigurable to accept new techniques. Within the existing channel rate, the slot structure can be redefined to support lower bit rates or variable bit rates. A modification, if carried out with careful attention to other architectural constraints, could be implementable by means of read-only memory (ROM) changes in the digital circuitry. Existing radio hardware can most likely be utilised at the base station; also, a TDMA format can be designed to accommodate different bit rates and different slot lengths. Since the channel rate, bandwidth, and other features of the radio transmission remain the same, such changes can be introduced without disrupting the cellular network frequency plan.

8.3 The pan-European cellular system

The pan-European cellular mobile radio system was conceived in 1982 by a committee of the *Conference of European Posts and Telecommunications Administrations* (CEPT). CEPT foresaw a need during the 1980s and 1990s for public cellular radio, but not accompanied by a widening divergence of systems, so that a standard was gradually put in place to positively encourage convergence. It also wished to achieve a measure of compatibility with ISDN. The system that emerged from this process has become widely known as the GSM system. GSM now stands for *global system for mobile communications;* previously it was named after the planning group shown in Figure 8.4. In 1989 GSM was transferred from CEPT to the *European Telecommunications Standards Institute* (ETSI).

ETSI is a legal entity, has over 289 members among the industrial community as well as the telecommunications operators, and a large permanent staff located in purpose-built accommodation near Nice, in the south of France. It deals with a wide range of telecommunication standardisation matters through a General Assembly and a Technical Assembly which meet several times a year; the relevant part of the structure of ETSI is shown in Figure 8.4.

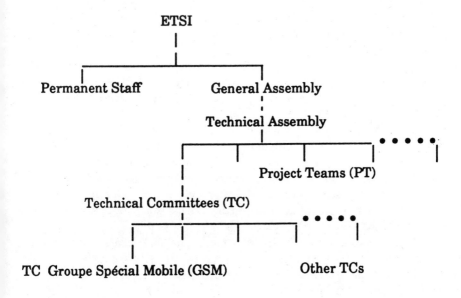

Figure 8.4   Outline structure of ETSI. The GSM TC is now known as the Special Mobile Group

8.3.1   Features of GSM

The primary function of GSM is to provide a full roaming mobile telephony service.

The services provided by the GSM system are recognised as falling into three categories:

- teleservices,
- bearer services,
- supplementary services

The term *teleservice* refers to those services provided on a user-terminal to user-terminal basis. The most important teleservice is clearly voice communication. Facsimile transmission also belongs to this category. Another example is the short-message service (a form of alpha-numeric paging) in which a message received from a mobile can be read directly from the built-in display.

In *bearer services* the terminal equipment is provided by the user, the responsibility of the network-service provider ending at the point of connection. Many forms of data transmission, at rates between 300 and 9600 bps, fall into this category. GSM provides no specific error protection for so-called transparent data links: in such situations the user must provide any necessary protection. In non-transparent data services, a GSM radiolink protocol will protect the data, but at the same time reduce the maximum data rate to 4800 bps.

Supplementary services will be developed along the lines of planned ISDN services, but are likely to vary greatly from country to country. Among the first to be introduced will be call forwarding, advice of charge, call barring and conference facilities.

With GSM, subscription is recorded in a subscriber card or 'smart card'. These are like a normal size credit card (or it may be a much smaller card for use in handheld phones), but contain a complete microcomputer with memory. Therefore these cards can perform many functions and also provide a high level of security. When such a card is plugged into a GSM phone it assigns the MIN code; it also checks that the subscription is valid, and the card not stolen, by authenticating the call back to the user's home database.

This provides exceptional security, preventing false charges to the user's account and ensures that incoming calls are correctly delivered. Other useful and novel features include the ability to store user information such as a list of short codes for dialling frequently used numbers. Voice security is also greatly enhanced by the use of full digital encryption. This applies equally for voice and for data calls.

Speech quality on GSM is comparable with analog systems under average to good conditions. However, under conditions of a weak signal or bad interference, GSM should perform significantly better.

The data services can offer a performance with lower errors, which will also be faster than presently available.

Size, weight and battery life are also important parameters of performance. Due to the digital circuitry employed, a high level of silicon implementation is be expected, leading to smaller, lighter phones as technology progresses.

The GSM recommendations fill some 5000 pages divided into the following main sections:

0	Preamble
1	General vocabulary, abbreviations
2.	Service aspects
3	Network
4	MS-BS interface and protocols
5	Physical layer on radio path
6	Audio aspects
7	Terminal adapters for mobiles
8	BTS/BSC and BSC/MSC interfaces
9	Network interworking
10	Service interworking
11	Network management, operations and maintenance
12	Equipment specifications and type-approval

Clearly the recommendations occupy a much greater area of paper than this book; however, because we are concentrating mainly on the cellular radio aspects, the critical recommendations can be summarised, by focusing on numbers 4 and 5 only, for example.

8.3.2 The OSI reference model

A system of the complexity of GSM, or indeed any of the other competing digital cellular networks, requires much planning and organisation, both in the definition and in its practical implementation. A pattern for structuring data communication networks in general has been provided by the *International Standards Organisation* (ISO) in the form of the *Open Systems Interconnection* (OSI) model. The OSI model provides for a number of horizontal layers, each layer communicating exclusively, and according to well-defined rules, with the layers immediately above and below it. Communication thus becomes vertical, rather than horizontal, except for the lowest, or physical layer, where the information is passed from one system to the other. The GSM specifications or recommendations as listed above have been written to fully define the lower three layers of this OSI model, see Table 8.2.

Table 8.2   OSI model and the use of three lowest layers (1 to 3) in GSM

OSI LAYER NO	OSI LAYER NAME	GSM EQUIVALENT MODEL	TASKS
7	Application	-	User tasks
6	Presentation	-	User tasks
5	Session	-	Network tasks
4	Transport	-	Network tasks
3	Network	call management mobility management radio resourse	GSM network tasks
2	Data link	concatenation segmentation acknowledgement	GSM network tasks
1	Physical	error detection channel coding modulation	GSM network tasks

- In the lowest layer, layer 1, the *physical* characteristics of the transmission or radio path medium are specified. In the context of the GSM radio link, this definition includes not only the frequencies, modulation types, etc., but also the structure of the bursts and frames implicit in a time division multiplex transmission scheme. Since this layer is responsible for the correct transmission of single bits, an element of error-protection coding also belongs here.

- Layer 2, or the data-link layer, consists of an intelligent entity responsible for the safe communication of meaningful messages or frames between radio stations. To this end the transmitting section structures the messages of the higher layer to match the physical constraints of the layer 1 medium and requests, in many situations, a confirmation (acknowledgement) from the receiving end. At the receiving end messages are reconstructed from the received frames and the acknowledgements are formulated for retransmission.

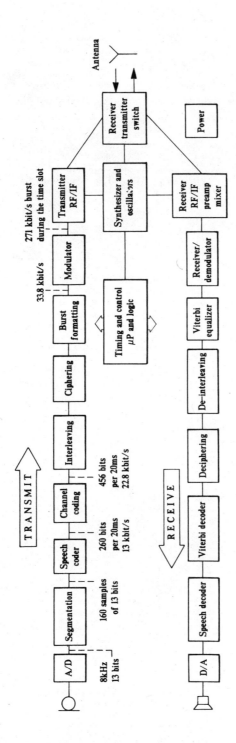

Figure 8.5   Functional block diagram of a GSM digital cellular mobile station

- Layer 3, the network layer, is responsible for management of all calling and related activity of the radio network. These tasks are further subdivided into sub-layers designated call control management, mobility management and radio resource management.

- The higher layers apply to any telecommunications system and specific reference to these tasks does not appear in the table of GSM recommendations, listed in section 8.3.1.

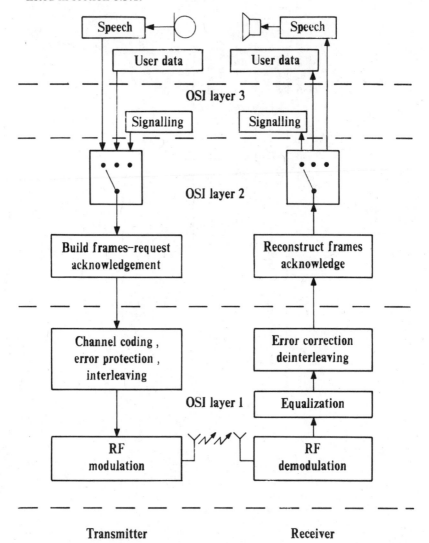

Figure 8.6   An OSI model for the mobile part, showing the first three layers

To illustrate how the OSI rules change the system operation into vertical communication layers, as opposed to the usual horizontal serial operation, Figure 8.5 outlines a GSM digital cellular mobile unit in the conventional manner. (This diagram should be compared to Figure 1.3.) The receive/transmit frequencies are similar, as could be the front-end circuits, except that digital cellular works in a burst mode. The back-end of the MS differs considerably, however, because of the massive digitization and control needed.

Drawn in the OSI format the same circuit functions can appear as shown in Figure 8.6.

A summary overview of the critical features which apply to GSM, as compared to ADC (or JDC), is shown in Figure 8.7. The differences are really those of *parameter values*: hence much of the detailed description (or understanding) of one can be transferred to the other, and this is assumed in the description that follows.

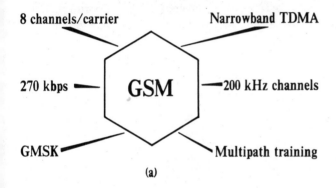

(a)

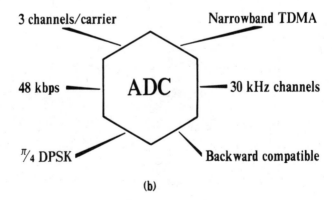

(b)

Figure 8.7   Features of narrowband TDMA digital cellular radio, (a) refers to GSM, (b) refers to ADC

8.3.3 The fixed network supporting GSM

When the GSM committee started its work much of the effort was concentrated on the radio system. This reflects the essential, and indeed only real difference between a mobile network and a fixed public switched telecommunication network (PSTN); that is, the substitution of the subscriber's local loop by a mobile radio path. It is therefore appropriate that attention is focused on the radio aspects; but, in order to illustrate the similarities to fixed networks, the network aspects of the system also need to be considered.

There are several interfaces; they are all important for the operation of the system, but only those relating to the radio and network access are discussed in any detail here. Other interfaces include, for example, the *subscribers interface module* (SIM), which personalises a mobile station. Some MSs will have a slide-in, smart-card SIM and some a wired-in one that is not user-changeable.

First of all, the fixed network is not unlike other networks supporting cellular; in fact, Figure 1.8 is very applicable to GSM. Within this architecture the interfaces can be defined as set out in Figure 8.8.

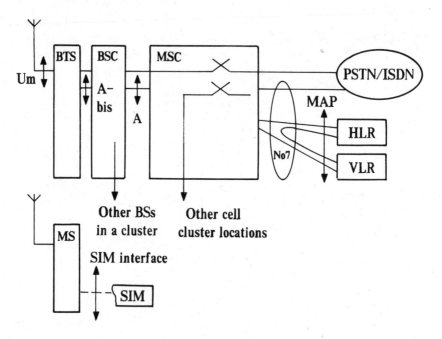

Figure 8.8   Principal interfaces within GSM

Two mandatory interfaces are (i) the A-interface which connects a *base station system* (BSS) to the MSC and (ii) the Um, or air interface, which provides the connection of the BS to the antenna, and hence MSs. Within the BSS part there is the *base station controller* (BSC) and the *base transceiver station* (BTS). Here the PSTN PCM digital speech can be converted to GSM RELP coded speech and vice versa; a less well defined A-bis interface is implied.

Referring to this diagram (or Figure 1.8, or indeed Figure 3.20, which described analog cellular), GSM will have at least one home location register (HLR), where all management data relating to all home MSs is stored, and at least one visitor location register (VLR), where selected data relating to visiting MSs is stored. This data includes the *international mobile station identity* (IMSI), the MS international ISDN number, and other information, including the current location of the MS. The HLR and VLR may be co-located with the mobile services switching centre (MSC), or not, as the system operator decides and, of course, the number of HLRs and VLRs will depend on the size of the network. Thus the interfaces between the MSC, the HLR and the VLR must also be defined, as indicated in Figure 8.8. MAP stands for the communication protocol between the network sub-system elements and is known as the *mobile application part* of GSM. Signalling system No. 7 is known as the *telephone user part.*

Every GSM mobile will be allocated to a specific HLR (the code of which forms part of the IMSI) and this information is used to enable calls to be made when, for example, a MS visits a network other than its home network. The MS is requested to pass the IMSI on the up-link to the nearest base station (BS) and this is forwarded via the MSC to the VLR of that network. The VLR accesses the HLR (via fixed network links, international if necessary) in order to obtain the selected information needed for registration.

A *MS roaming number* (MSRN) and a *temporary mobile subscriber identity* (TMSI) are allocated by the visited network. An authentication process is carried out; if this is successful access is permitted and MS-originated calls can then be made, while the registration of the MS's location enables incoming calls to be routed to the correct area.

Among the selected data transferred from the HLR to the VLR will be the services available to that MS, for example, supplementary voice services or data services. At the time of initiating a call, specific details of the services requested will then be required from the MS. It can be seen that the signalling activity on both the radio path and on the fixed links of the system is considerable. In fact, emulation of digital cellular networks shows that the utilization rate of the HLR to VLR path increases faster than any other link in the overall network and this rate may limit the dimensioning of an installed network.

The international signalling system CCITT No. 7, as used by the fixed telephone services, has been specified for the network management and interconnection signalling functions of the GSM system; the requisite mobile application part is in fact one of the largest of the GSM recommendations and runs to nearly 600 pages. This illustrates the level to which GSM standardisation is being implemented by the ETSI GSM technical committee.

8.3.4   The radio part

The radio part is known as the *radio sub-system* (RSS) in GSM and comprises the various aspects of the interface between base stations and mobile stations - the mobile radio link. Often called the *air interface*, it includes the definitions of the logical, i.e. the traffic and control channels; the physical channels, i.e. the radio frequencies and time slots, the multiple access, multiplexing and time slot structures, frequency hopping, coding and interleaving, modulation, power control and handover, synchronisation and transmission and reception.

Also included are the specifications of spurious emissions which could potentially have an adverse effect on other services. These arise from the modulation process and the power ramping at the beginning and end of each time slot, as discussed below, as well as the usual imperfections in the actual implementation of the oscillator, frequency synthesizer and other radio frequency elements that make up a mobile radio station.

All the physical (layer 1) radio aspects are covered in the RSS recommendations and are accommodated within fewer than 200 pages, whereas the layer 3 functions relating to the MS-BS interface require many hundreds of pages for their full specification.

Like the first generation TACS network, several classes of mobile are assumed. A class 1 MS (typically vehicle-mounted) has a peak power output of 20 W or 43 dBm, whereas a class 4 hand-held portable has a peak power of 5 W (37 dBm) and a class 5 only 2 W (33 dBm). The maximum receiver sensitivity of both BSs and vehicle-mounted MSs is −104 dBm; that of hand-held portables is −102 dBm. Hence it can be seen that when the path loss reaches a limiting value for the hand-held portable it is at a disadvantage compared with vehicle-mounted MSs which still have several dB in hand.

This can be expected to occur most often in rural areas where the cells are typically large and the coverage can become patchy. In these situations hand-held portable MSs may not give such good service as the more powerful MSs in classes 1-3, but they will perform excellently in the small cells that typify dense urban and city areas and where there will be much greater numbers of users. In such city areas with a large number of small cells the maximum BS power output is unlikely to be used, and while this does not balance the path loss, it ensures that hand-held MSs are used well within their capabilities in most cases. The exception to this is the use of hand-held MSs inside buildings, lifts, underground railway stations and other situations where radio penetration from a conventional mast top BS antenna is weakened, as discussed in Chapter 2.

Table 8.3 summarizes several of the parameters of the air-interface. Some of these were referred to in Table 8.1. In the context of the discussion here, the two critical parameters are the R_X/T_X spacing. In frequency, it is 45 MHz (as is the case for TACS); in time, it is three time slots. Thus, in Figure 8.2 the reverse channel is operating three time slots later.

Table 8.3   GSM air-interface parameters

Mobile to BS	890-915 MHz (note Table A.1)
BS to mobile	935-960 MHz
Channel spacing	200 kHz
R_X/T_X spacing - frequency	45 MHz
R_X/T_X spacing - time	1.15 ms
Modulation	0.3 GMSK
Users per frequency pair	8 initially (16 with $\frac{1}{2}$ rate coder)
Frame period	4.615 ms
Time slot period	576.9 µs
Symbol period	3.692 µs
Symbols/time slot	148

A good way of illustrating the radio paths involved is shown in Figure 8.9. A BS is assigned to each channel group, but can communicate to eight MSs since there are 8 time slots, see Figure 8.2. Meanwhile, the MS has to switch from R_X to T_X in about one msec (2 time slots), but more time is allowed for T_X to R_X operation. Other units will be operating in other channels or time slots.

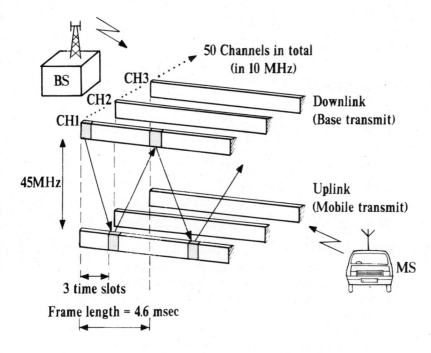

Figure 8.9   RF activity between a base station and a mobile

8.3.5   The timing structure of GSM

The most prominent characteristic of the GSM physical layer is the elaborate
timing structure. GSM carriers are spaced at 200 kHz intervals, each carrying a
270.833 kbps digital signal. To organise the information transmitted on each
carrier, GSM defines several time intervals, ranging from 0.9 μs (one-quarter bit)
to 3 h 28 m 53.760 s (*cryptographic hyperframe*).

Terminals and base stations insert information into the channel in *time slots*,
each of duration 577 μs. Eight consecutive time slots comprise a TDMA *frame* of
duration 4.62 ms, while 26 frames comprise a *multiframe* with duration 120 ms.
Figure 8.3 introduced the arrangement of a timeslot in a frame, now shown in the
lower part of the diagram of Figure 8.10.

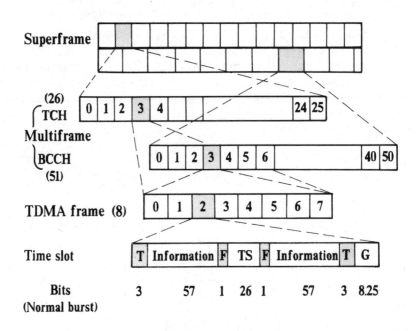

Figure 8.10   The TDMA timing organisation heirarchy within GSM

We deal firstly with the *traffic channels* shown here. We will return to the
broadcast control channels (BCCH) later.

In GSM the speech, or data traffic, is not placed packet-by-packet in successive
time slots. Channel coding is used to arrange error-protected speech data into the
final form necessary for RF transmission. Channel coding involves adding additional data for channel control, training sequences, and tail/guard bits, shown in
Figure 8.11. In addition, the channel coder must interleave the data to enhance the

performance of the error correction and rearrange the data into packets for transmission. Training sequence data is added to the voice data which aids in data identification and is also used for equalization of the RF channel. The tail/guard bits provide a buffer between adjacent data packets. Once additional control channel data has been added by the channel coder, the data is interleaved and arranged into packets for RF transmission (for a TDMA system). It should be noted that after channel coding, the additional data increases the overall data rate. To allow multiple users to share a physical channel, the data is compressed in time and output at a much higher data rate, see Figure 8.5.

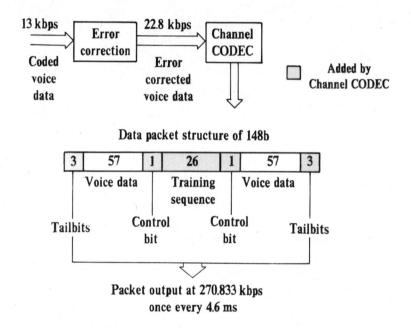

Data packet structure of 148b

Packet output at 270.833 kbps
once every 4.6 ms

Figure 8.11   Channel coding arrangement for introducing a burst of coded voice data into a time slot

Figure 8.11 shows that the error-correction unit increases the bit rate to 22.8 kbps, which was discussed in the last chapter (Figure 7.7). Use of the *channel coder* unit, shown in Figure 8.12, puts the traffic data in 20 msec blocks of 456 b each.

As just described, the 22.8 kbps data is applied to the channel which arranges this data into packets of 148 b. The two blocks of 57 bits of voice data are also interleaved by the channel coder to spread out the voice data over many packets. Each packet is then outputted at the much higher data rate of 270.833 kbps, but a packet (length 547 μsec) is only sent once every 4.6 ms. This is done so that voice data from 8 users can share a physical channel in a TDMA system. Each user sends a packet which uses 1/8th of the total time available on the 270.833 kbps data stream. Once this data has been processed according to Figure 8.12 it is ready to be transmitted via the RF carrier. We return to this strategy again below.

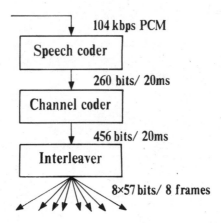

Figure 8.12   Arrangement for interleaving 57 b bursts of coded voice data over eight frames to make up the 20 msec speech burst

Returning to the multiframe shown in Figure 8.10, which has 26 frames, 24 frames are user information. The other two frames carry system control information. A frame consists of eight slots, with each slot assigned to a different mobile terminal.

To zoom in on a time slot, we divide 120 ms (multiframe duration) by 26 (frames per multiframe) to obtain 4.62 ms per frame. With the frame divided into eight slots, the slot duration is 577 μs, or 156.25 channel bits, including a guard time of 30.5 μs (8.25 b). No energy is transmitted in the guard interval, which allows the system to operate with unpredictable mobile-to-base arrival times.

The composition of a GSM time slot and its placement in a frame and multiframe were shown in Figures 8.10 and 8.11. Two user information bursts of 58 b account for most of the transmission time in a slot. (57 b carry user information, while the other bit is used to distinguish speech from other transmissions.) There is a synchronization burst consisting of 26 b in the middle of the time slot, and the slot begins and ends with three tail bits, all logical zeros.

The GSM modulation is Gaussian minimum shift keying, as described in Chapter 6, in which the modulator bandpass filter has a 3-dB cutoff frequency of 81.25 kHz (0.3 times the bit rate). The modulation efficiency of 271 kbps operating within a 200 kHz carrier spacing, is 1.35 bps/Hz. With a bit interval of 3.7 μs, which exceeds typical delay spreads (see Chapter 5), the GSM signal will encounter significant intersymbol interference in the mobile radio multipath propagation environment. As a consequence, an important component of a GSM receiver is the adaptive equalizer necessary to provide reliable binary signal detection, apart from the channel coding strategies.

8.3.6   Channel coding and training sequence

The channel coding technique chosen for the encoded speech (see Chapter 7) employs a convolution code of rate 1/2, block-diagonally interleaved over 8 TDMA frames to provide protection against burst errors, shown in Figure 8.12.

The 20 ms speech frame consists of 260 bits; a 3 bit cyclic redundancy check (CRC) is added to the 182 most significant bits making 185 bits; 4 tail bits are then added (to initialise the decoder) to produce 189 bits which are then presented to the 1/2 rate channel encoder producing 378 bits. The remaining 78 least significant coded speech bits, which are unprotected, are then added making a total of 456 bits (see Figure 7.7).

The bit rate of the resultant protected channel is therefore somewhat higher than that of the speech encoder output at 22.8 kbps, but the traffic channel is now fairly robust and is capable of providing good communication under adverse conditions.

The effect of errors on the signalling, system control messages and data are not the same as for speech, so different coding rates are employed in order to provide a balanced degree of protection.

In the GSM system the long excess delays resulting from signal components, whose paths are very long, result in a measurable delay spread which can only be tolerated by using an equalizer in the receivers.

A bit pattern, known as the *training sequence* (26 bits), TS in Figure 8.10, is transmitted at regular intervals, and the equalizer compares the received bit pattern with the training sequence, adjusting the parameters of a digital filter so as to produce the inverse transfer function to that of the radio path. As the changes are rapid, and because the time slots are independent, every time slot must contain the training sequence.

This is an extra overhead on the information-carrying capacity of the digital channel and further reduces the effective throughput. Without equalization, however, the effective range would be limited to perhaps a few hundred metres; on the other hand, a cheaper mobile station with no equalizer is conceptually possible for very short range operation, e.g. microcell operation.

In the GSM system the design of the equalizer is not specified and is left to each manufacturer, but its performance is defined in terms of a maximum excess delay of 16 μs. The definition is quite complex and is set by type approval conditions.

Type approval is subject to conformance with a set of tests using a system simulator defined in the recommendations. This makes use of a propagation simulator which models a number of multipath profiles together with appropriate Doppler spectra. The models emulate urban, rural and hilly terrain conditions at specified vehicle speeds. A vehicle is anything which physically transports the mobile, including a pedestrian, thus causing it to move in relation to a base station and thereby creating a non-stationary signal reception pattern. The equalizer must perform satisfactorily so that the resultant error performance is within specified tolerances during a type approval test.

8.3.7   Radio link management

The radio link, comprising both uplink and downlink, has to be managed in order to ensure continuity of service and to minimise interference to other users of the system. This is a complex topic but the main points will be addressed.

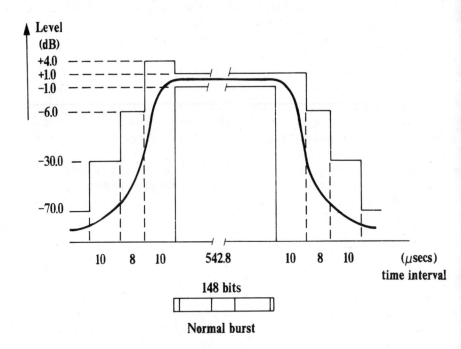

Figure 8.13   GSM transmitter power ramp specification

Timing advance: the propagation time of the radio signals between the MS and the BS will be determined by the distance that the radio signal travels. Since cells in the GSM system may vary in size from perhaps 1-2 km up to about 35 km, it can be seen that the propagation time can vary from about 3-100 microseconds. In order that the data bursts transmitted by each MS fall exactly into the time slot structure at the BS receiver it is necessary to advance the timing of the MS transmitter by an appropriate amount and this must be done individually for each MS. This is done at call set-up by the BS which measures the round trip time and sends a timing advance message to the MS, which is amended as necessary at intervals during operation.

Power ramping: the power burst of all GSM transmitters must be very tightly controlled. Unlike analog cellular where the transmitter mask specification relates to power versus frequency (see Figure 8.1), in GSM a very specific power ramp-up and power ramp-down mask has to be specified for just one time slot of 148 b occupying 542.8 µsec. This mask is shown in Figure 8.13.

This mask ignores the 8.25 guard bits, which are built-in for the timing advance procedure, and also loses half of the first and last tail bits so in effect there are only 147 useful bits. This specification has to be met at the same time as the power frequency mask (Figure 8.1), which is a function of the GMSK modulator performance.

Adaptive power control: this is applied to all MSs to ensure that they operate at the lowest power level consistent with adequate received signal strength and quality. The power is controlled in steps of 2 dB from the maximum defined by the power class.

For example, a class 1 MS (peak power 20 W or 43 dBm) has 15 steps giving a minimum power level of 13 dBm. The other classes are catered for by the same algorithm since their peak powers, 8 W (39 dBm), 5 W (37 dBm) and 2 W (33 dBm), all correspond to steps on the same scale.

Power control is achieved by a process which involves measurement, by the MS, of the received signal strength, quality and the regular reporting, on the uplink, of these data to the BS. The BS has preset parameters to enable a decision to be made and, when a threshold is reached, it commands the MS to change power level either upwards or downwards as necessary. The received quality is a measure of the BER.

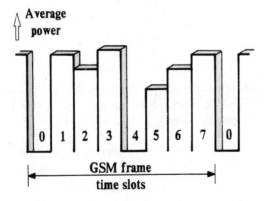

Figure 8.14   Possible power profile for a GSM base station

Base stations may have an adjustable peak power level to allow the system operator to make adjustments to the cell coverage area and in addition they may employ power control on the downlink in a similar way to that defined for MSs. This makes the power profile of any BS complex over successive GSM frames, an example being given in Figure 8.14. Each component of the profile must meet the specification of Figure 8.13.

Handover: this is one of the basic features of cellular radio and enables MSs to move freely across cell boundaries and enjoy continuous service. It uses the same process of measurement and reporting as for power control; indeed the two procedures are closely linked.

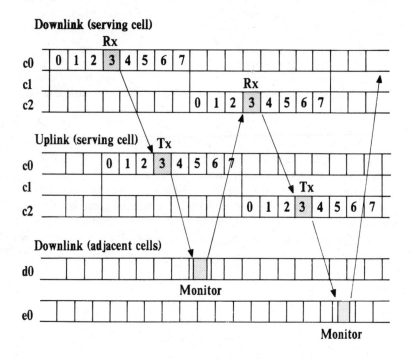

Figure 8.15   Mobile station usage of channels during a call, showing monitoring of adjacent cell broadcast control channel time slots, and frequency hopping if used

The GSM system exploits the properties of TDMA very effectively for handover; the MS listens for the BCCH of up to 16 surrounding BSs in the time slots that are not being used for transmission or reception, see Figure 8.15. It forms a list of up to 6 handover candidates and reports the signal strength and quality to the serving BS. Meanwhile the BS is also monitoring the signal strength and quality on the serving channel. Handover is under the control of the network, usually the base station controllers, and is used to provide continuity of communication by the availability of another channel which can allow communication at a lower power level. Handover can also be used for traffic balancing between cells.

Handover to another cell will involve retuning to another radio frequency channel, but handover is also possible to a different time slot on the same radio frequency channel in the same cell, which may be used for interference control reasons. In each case one has mobile assisted handover.

8.3.8 Signalling within GSM

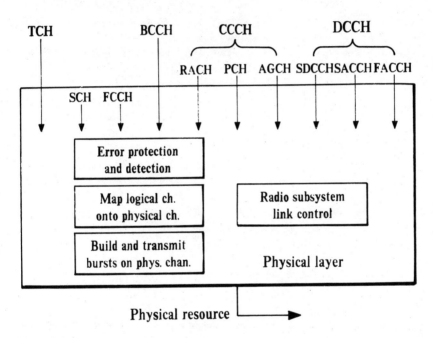

Figure 8.16   Mapping of the logical channels onto the physical RF channel

There are several types of control channel. The main ones are: a *broadcast control channel* (BCCH) on downlink only, to keep MSs aware of the base station identity, frequency allocation and frequency hopping, if used; a *common control channel* (CCCH), further subdivided to provide a *random access channel* (RACH) on the uplink; a *paging channel* (PCH) and an *access grant channel* (AGCH) on the downlink; a *dedicated control channel* (DCCH) for registration, location updating, authentication and call set-up; two *associated control channels* (ACCH), one a continuous stream *slow* ACCH (SACCH) for call supervision and burst stealing mode; one a *fast* ACCH (FACCH) for power control and handover. These channels are logically mapped onto the physical RF channel, as shown in Figure 8.16.

The SACCH is carried in one frame of every 120 ms multiframe. Each active terminal owns a slot in the SACCH frame. A SACCH message contains 184 information bits. Error control coders process this information to produce 456 channel bits which are interleaved and distributed over four time slots. With one time slot per multiframe, and a multiframe duration of 120 ms, the total transmission time of a SACCH message is 480 ms. Thus the transmission bit rate is 456 b in 480 ms, or 950 bps. The information rate is 184 b in 480 ms or 383 bps.

When more rapid network control is required, GSM creates a *fast associated control channel* (FACCH) by interrupting user information for a duration of four frames. The delivery time of a FACCH message is four frames, or 18.5 ms, in contrast to 480 ms in the SACCH. The contents of bits in each slot, and slots in the basic TDMA frame are shown in Figure 8.17.

The BCCH is organised into a 51-frame multiframe as shown earlier in Figure 8.10, and is carried in time slot zero on a non-hopping radio frequency carrier; the remaining time slots are available for traffic channels. Several radio frequency carriers are allocated to a BS, of which one will carry the BCCH. The use of hopping or non-hopping radio frequency carriers for the traffic channels is optional.

Unlike analog cellular the traffic messages and the signalling messages are part and parcel of the same frame make-up arrangement within GSM. It is therefore perhaps useful to go over these time slots and channels, as they are called, again so that the correspondence to analog cellular becomes more obvious.

- Traffic channels are used to carry digitized speech or other data. The traffic channels are further classified by the speed of transmission possible. Speech transmissions are defined as

    - full-rate traffic channels (22.8 kbps) and, in the future
    - half-rate traffic channels (11.4 kbps)

To carry user data a subscriber may choose again between half-rate and full-rate traffic channels with data rates of 2.4 and up to 9.6 kbps.

- Control channels carry signalling or synchronizataion data. There are three different types of control channels and special channels within these categories as follows:

- Broadcast channels carry information for frequency correction of the mobile stations (FCCH) and provide them with the possibility for frame synchronization (SCH). After identification and synchronization the mobile stations use the broadcast control channel (BCCH) to get more information about the base station.

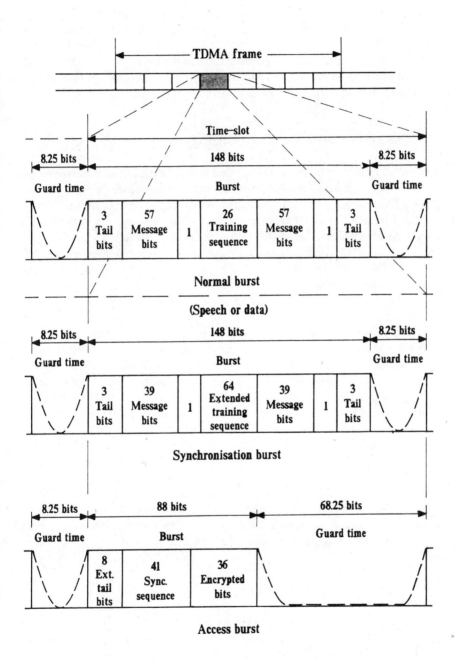

Figure 8.17   Three of the time slot bursts in the GSM system

- Common control channels are used to organize the initial commands for establishing a call before changing to a special 'dedicated' control channel.

- Dedicated control channels are used for information interchange between BS and MS before a call, using a so-called *stand alone dedicated control channel* (SDCCH) and during established calls (SACCH/FACCH).

8.4   The North American digital cellular system

The digital cellular radio system developed for North America differs from GSM in many ways. Two underlying reasons for the differences are found in the most pressing requirements of each system and in their regulatory constraints. While a principal purpose of GSM is to establish, for the first time, a unified public telephone system for all Europe, the United States and Canada have had the benefit of a single standard (AMPS, Advanced Mobile Phone Service), since the introduction of cellular services in the early 1980s. While GSM was designed as a standalone system operating in newly assigned frequency bands, the new North American digital transmission techniques must co-exist with present day analog systems in the established cellular frequency bands.

The development of the North American digital transmission standard comes at a time of high demand for cellular services with no new spectrum available to meet this demand. Therefore, the overriding aim of the new technology is to increase the capacity of the existing spectrum to provide increased services. In 1988, encouraged by the *Cellular Telecommunication Industry Association* (CTIA), consisting mainly of cellular service providers, the *Telecommunication Industry Association* (TIA) of equipment manufacturers established a technical committee to develop a digital standard. Like GSM, the TIA stimulated the production of prototype equipment, which was then subjected to field trials (access and modulation technologies) and laboratory tests (speech coders). In 1989, the industry, by a majority vote, adopted specific aspects of the dual-mode system, and in 1990 it accepted the entire transmission standard which is referred to as the Electronics Industry Association Interim Standard 54. Hence, the name IS-54, which, at this time, is the most common nomenclature for the new technology.

8.4.1   Radio transmission strategy

The carrier spacing of IS-54 is 30 kHz, as in the first generation AMPS system. Operating companies will selectively convert analog channels to digital operation in order to relieve traffic congestion at cellular base stations. Each digital channel operates at 48.6 kbps, carrying three user signals. The ADC air-interface parameters are shown in Table 8.4, which should be compared with GSM in Table 8.3.

Table 8.4   ADC air-interface parameters

Mobile to BS	824-849 MHz
BS to mobile	869-894 MHz
Channel spacing	30 kHz
R_X/T_X spacing - frequency	45 MHz
R_X/T_X spacing - time	1.85 ms
Modulation	$\pi/4$ DPSK
Users per frequency pair	3 initially (6 with $\frac{1}{2}$ rate coder)
Frame period	40 ms
Time slot period	6.6667 ms
Symbol period	41.15 µs
Symbols/time slot	162

The frame/time slot structure of ADC is shown in Figure 8.18. Note how control and traffic data is now combined within each time slot.

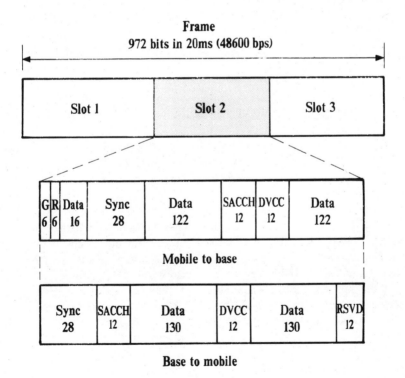

Figure 8.18   IS-54 time slot and frame structure (for definitions see text)

One time slot carries 324 b including 260 b of user information and 12 b = SACCH of system control information. The remaining 52 b carry a *time synchronization signal* 28 b (= SYNC), a *digital verification colour code* 12 b (= G) and, in the mobile-to-base direction, a 6 b guard time interval, when no energy is transmitted, followed by a 6 b (= R) ramp interval to allow the transmitter to reach its full output power level. 12 b (= RSVD) are reserved.

The 28 b synchronization field contains a known bit pattern that allows the receiver to establish bit synchronization and to train an adaptive equalizer. The system specifies six different synchronization patterns, one for each slot in the 40 ms frame. This allows the receiver to lock onto its assigned time slots. The digital verification colour code plays the role of the supervisory audio tone in the analog AMPS system. There are 256 (8) b colour codes, protected by a (12;8;3) Hamming code. Each base station is assigned one of these codes and the verification procedure prevents a receiver from locking onto an interfering signal from a distant cell.

The IS-54 speech coder, described in Chapter 7 as a codebook vocoder and, as shown in Figure 7.10, processes input signals in blocks of duration 20 ms, just as in GSM. Also, in common with GSM, the transmitter assembles 40 ms of speech information and interleaves the bits in order to spread short bursts of channel errors over a longer time interval. Because of the different coding rate and timing structure, 40 ms of speech information (520 b) goes into two IS-54 time slots, rather than the eight slots of GSM.

In a departure from the constant envelope modulations of first generation cellular radio systems and other second generation systems, ADC adopts a linear modulation technique, namely, differential phase shift keying (DPSK), as described in Chapter 6.

The transmission rate is 48.6 kbps with a channel spacing of 30 kHz, which comes to 1.62 bps Hz, a 20% improvement over GSM. The main penalty of linear modulation is power efficiency, which affects the weight of hand-held portable terminals and time between battery chargings. The specific method of modulation is $\pi/4$ shifted DPSK, with root-cosine rolloff filtering at the transmitter and receiver. The rolloff factor is 0.35, which provides a spectral zero in the transmitted signal at 16.4 kHz. It is also of interest to note that the highly planned *airborne public correspondence* (APC) service for Europe is also recommending the use of the $\pi/4$ shifted DPSK, and a very similar narrowband TDMA format, and the use off a multi-pulse exited LPC speech encoder. The service will be known as the *terrestrial flight telephone system* (TFTS). APC at present operates in the USA in the conventional FDMA FM format.

8.4.2   Control channels

As a dual-mode system, in the sense that both analog and digital MSs can have access to the system, IS-54 retains much of the control structure of the original

AMPS system. Call set-up is provided by forward control channels and reverse control channels, which carry messages between base stations and mobile telephones. Augmenting the original message set is an indication that a mobile terminal has dual-mode capability. When a call is established with a dual-mode terminal, the system has the option of assigning this call to a digital channel. The system also has the capability to handover a dual-mode mobile from an analog voice channel to a digital channel, and vice versa. When it operates over a digital channel, the dual-mode terminal exchanges messages with the system through a slow associated control channel (SACCH) and a fast associated control channel (FACCH), as in GSM. The control channel messages have a length of 65 b, and the FACCH in IS-54 operates in a way similar to its counterpart in GSM, by interrupting user information to send urgent control messages. In the FACCH, forward error-correction is provided by a convolutional coder, that produces 260 b which replace the user information in one time slot. In the case of the SACCH, there is a half-rate convolutional coder that generates 132 b per message. This information is interleaved over 12 time slots and transmitted in the SACCH fields shown in Figure 8.18. The total delay of the SACCH is six IS-54 frames, or 240 ms.

A similar cellular radio system to ADC is a Japanese digital cellular network known as JDC. Like the North American scheme it is intended to overlay their present analog cellular. Since this is a 25 kHz spacing, the bit rate has been reduced to 42 kbps. Again using $\pi/4$ DPSK, a modulation efficiency of 1.68 bps Hz is aimed at, which is the most efficient planned so far. A VSELP speech coding algorithm working at 11.2 kbps is also being adopted.

Further reading

Alcatel SEL. (1989) 'System 900 Mobile Cellular Radio', *Electrical Communications*, 63, Number 4, pp 383–414

Balston, D. M. (1989). 'Pan–European Cellular Radio, *IEE Elec & Comms J*, No. 1, pp 7–13

Calhoun, G. (1988). *Digital Cellular Radio*, Artech House, NY

Cheesman, D.S. (1991). 'The pan-European cellular mobile radio system', in *Personal and Mobile Radio Systems*, Peter Peregrinus Ltd, UK

D'Aria, G., Muratore, F. and Palestini, V. (1992). 'Simulator and performance of the pan-European land mobile radio system', *IEEE Trans Vech Tech*, VT-41, May, pp 177-189

Ericsson Radio Systems AB. (1991) *Cellular Mobile Telephone System, CME 20,* Stockholm, Sweden

Goodman, D.J. (1991). 'Second generation wireless information networks', *IEEE Trans Veh Tech,* VT- 40, May, pp 366-374

Hewlett-Packard Co (1992). *Communication Test Symposium -Digital RF Communications,* Feb, London, UK

Kinoshita, K., Kuramoto, M. and Nakajima, N. (1991). 'Development of TDMA Digital Cellular System based on Japanese Standard, *IEEE Veh Tech Conf,* May, pp 642–645

Lindell, F. and Raith, K.(1990). 'Introduction of Digital Cellular Systems in North America', *Ericsson Review,* Feb issue

Motorola Ltd. (1992). *An Introduction to the Pan-European Digital Cellular Network,* European Cellular Infrastructure Division, Swindon, UK

Pautet, M-B., and Monly, M. (1991). 'GSM Protocol Architecture: Radio Sub-System Signalling', *IEEE Veh Tech Conf,* May, pp 326–332

Raith, K. and Uddenfeldt, J. (1991). 'Capacity of digital cellular TDMA systems', *IEEE Trans Veh Tech ,* VT-40, May, pp 323-331

Rohde & Schwarz UK Ltd. (1992). *Testing for GSM; Seminar notes,* April, London

Smith, D.R. (1985). *Digital Transmission Systems,* Van Nostrand Reinhold, NY

9 Spectral Efficiency Considerations

9.1 Introduction

At the end of the day, whatever cellular radio system is considered, the following two factors will have to be taken into account:

(i)   The number of subscribers that can be connected to the system in an acceptable way for the radio spectrum allocated

and

(ii)   the cost of the network to provide the number of subscribers contemplated with a satisfactory service

The second factor is a subject of considerable interest to telecommunication operators, but it is a subject somewhat outside the scope of this text. Indeed it is a matter fundamental to all public telephone operation, but academic textbooks on the matter appear to be scarce.

Usually, much of the fixed network is in existence for the fixed subscribers, and therefore the cellular network operator, having been granted a licence from the government or state to operate a cellular radio network, needs to invest in:

Base stations and sites
Land lines (or microwave links) to switching centres
Mobile switching centres
Billing centre and invoice distribution
Network control centre
Management overhead (=staff)
Maintenance programmes

This assumes that the operator has access to the existing fixed network.

Notice that the cost of the mobile need not enter into the calculation, since provided that this can be kept low enough to attract subscribers, it need not concern the network operator.

Digital cellular radio is attractive because base station transmitters and receivers each serve several channels; eight in the case of GSM. Hence the network can be expanded at a lower cost, at least in principle.

Conversion equipment, from one speech coding method to another will in general be needed, but otherwise much of the signalling can be common to the existing fixed network and the fixed part of the mobile network. Also there is more opportunity to reconfigure the network as traffic conditions change, by means of digital control.

However, the matter of spectral efficiency is not reconfigurable. Once the limit marked as factor (i) above is reached, the subscriber base cannot be expanded and, more seriously, complaints of blocking or no line access and handover failures will arise. We review below the basics on which figures for spectral efficiency of cellular radio systems have been and are derived.

9.2 Network example

Meanwhile as an example, suppose one were to consider the setting up of a cellular radio service in a country where none existed, at least, when this work was begun. Bolivia is the example chosen and Figure 9.1 shows a geographical sketch of the country of Bolivia. (More recent information, September 1992, tells us that a single service by Telefonia Celular De Bolivia is in operation with about 800 subscribers, possibly, in the capital La Paz.)

The territory is interesting in the sense that much of it is a large flat plain. On the other hand much of the left-hand region has difficult and partly inaccesible mountains. (The main towns are marked by a letter code, for the pupose of this exercise.)

Clearly one would want to have the mobile switching centres in the main towns, and the coverage in each city would be planned on a hexagonal cell pattern, of sizes sufficient to meet likely subscriber density, as discussed below.

In the rural plain, area base station sites could be set out following a national grid plan, like Figure 3.8. This would be entirely suitable using large cell clusters, i.e. a twenty-one repeat pattern. Examination of the map shows that about 168 sites would be sufficient to cover the plains with 40-50 km radius radio cells.

Hence the number of switching offices required would be

$$\frac{\text{No. of BS}}{\text{cluster size}} = \frac{168}{21} = 8$$

This number one notes is not excessive and they could be placed in the main towns, except for two in the far west region where there appears to be no main towns. To have good coverage, a low UHF frequency allocation such as 450 MHz, would be a good choice. This very basic plan is sketched over the map of Figure 9.1.

Note, however, how a 21-cluster plan uses up bandwidth (or channels). Assuming a TACS specification and allowing only 4 message channels per cell, requires

$[4 + 1 \text{ (control)}] \times 25 \times 21$
and twice for duplex = 5.25 MHz

GEOGRAPHY OF BOLIVIA
(Reproduced from tourist map)

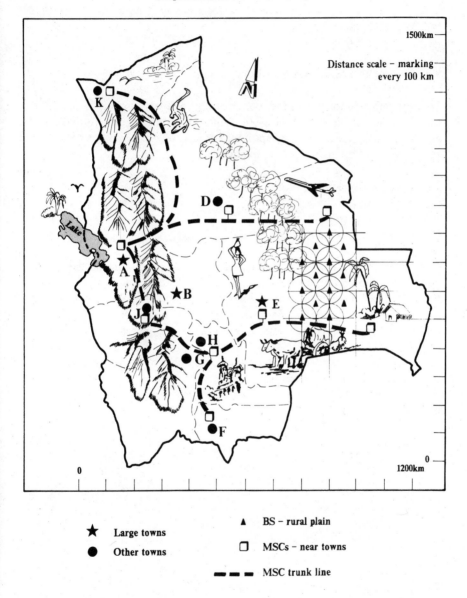

Figure 9.1   Outline geography of Bolivia and possible BS/MSC locations; Each base station would need 5 R_x / T_x plus one for standby, plus microwave link to main switching centre

The mountainous area is a much more difficult matter. Cell shapes would no longer be circular (or hexagonal). BS siting (and maintenance) would be difficult, but above all else where would the subscribers be? No general solution applies to this type of situation.

9.3   Bandwidth limit to subscribers

When it comes to towns or other conurbations, cell layout plans as described in Chapter 3 would be used. It is useful to derive a simple guide formula for the number of radio subscribers, in a province or city, who can expect to be offered a service at any particular time. For example let

Area of the city being considered	=	A km^2
Population of the city	=	P thousands of people
Average radius of radio cell	=	R km (see Figure 1.6)
Number of radio channels in each cell	=	n_c

n_c will depend on what cell repeat pattern is used and also on the spectrum allocated to the service. Cell patterns were described in Chapter 3; in particular Figure 3.13 will be useful, as will the associated discussion of channel allocation.

A radiotelephone system designed on a cellular basis also uses a dynamic assignment or trunking technique which can be interpreted as a channel gain, namely Figure 3.7, where typically 30 users think that they each have individual access to one private channel. Therefore, the number of users supported by each radio channel in the cell is multiplied by this factor.

We will use a factor of times 30; approximated as equal to the number 10π in the equation below.

$\therefore$ Number of users $\approx 30\,n_c \approx 10\pi n_c$

$$\text{Meanwhile, number of cells in city } = \frac{A}{\pi R^2} \text{ (uniform cells)}$$

$$\therefore \text{ Total number of users supported } \approx 10\,\pi n_c \times \frac{A}{\pi R^2}$$

$$\text{As a percentage of number of population } \approx \frac{10\,n_c A}{P \times 1000 \times R^2} \times 100$$

$$\text{or } \% = \frac{A n_c}{P R^2} \qquad \qquad \dots (9.1)$$

Example

Consider a city of about half a million people, within a radius of six kilometers

$$\therefore \quad P \approx 500,000 = 500$$
$$A \approx 120 \text{ km}^2$$

using $\quad R = 2$ km

and $\quad n_c = 40$

% offered service $= \dfrac{120 \times 40}{500 \times 4} = 2.5\%$ of the population

The result shows that highly populated cities (A/P ratio small) are the most difficult to serve; also the cell size is critical, whilst one must maximize the number of channels in the allocated radio spectrum.

The result can, however, be rewritten involving the system allocated bandwidth ΔF, the channel bandwidth Δf_a, and cell cluster size N, because

$$n_c = \frac{\Delta F}{\Delta f_a N} \qquad \qquad \ldots (9.2)$$

$$\therefore \quad \% = \frac{A \, \Delta F}{P \, R^2 \, \Delta f_a N} \qquad \qquad \ldots (9.3)$$

This equation given previously as (6.3), shows that the success of cellular arrangement depends on the spectrum allocated per user, i.e. the term Δf_a, which, as was discussed above in section 6.2, is greater than the signal bandwidth Δf_m.

The result also shows that for dense subscriber scenarios, a good allocation of spectrum is needed, plus small cells and small cell clusters, where possible.

As Figure 3.13 showed, the term N does not reduce easily. A C/I hurdle has to be overcome. The recent TDMA technologies appear to offer the opportunity of a reduction in N from 7 to 4.

9.4 Measures of spectral efficiency

Spectral efficiency for a land mobile radio system can be measured in two ways

(i) Voice channels per MHz per km^2

The number of radio channels per MHz of spectrum bandwidth is most easily appreciated for an FDMA system, for example Figure 3.1. The number of times that the same channels can be used in a given land area depends on the cluster arrangements.

This definition of spectral efficiency is easily calculated in general, since it is based on fully specified parameters, and we will proceed this way. However, spectral efficiency can also be defined in terms of the telephone traffic intensity, which can be supported by the network, giving the alternative definition

(ii) Erlangs per MHz per km^2

This measures the quantity of traffic on a voice channel, or group of voice channels, per unit time. It is useful to recall the definition of traffic intensity here.

9.4.1   Definition of traffic intensity

The study of *traffic* is a well-established discipline in telephone engineering. Note that the word traffic is not used to refer to the subscribers, even if they have radiotelephones.

Telephone calls are made by individual customers according to their living habit or conduct of their business. The aggregate of customers' calls follows a varying pattern throughout the day, and facilities sufficient in quantity to provide satisfactorily for the period of maximum demand, called the *busy hour,* is a matter for PTO planning. The basic factors involved in the design of facilities are the *call attempt rate, call-holding time, numbers of channels* (trunks or facilities), and *grade of service.*

The product of the first two factors is the *offered traffic*. It denotes the amount of time that a number of callers desire the use of facilities.

Thus, offered traffic   =   (attempt rate) × (call-holding times)

A load that engages one channel (trunk) completely is known as an Er*lang* (defined below). Offered traffic is also expressed in terms of hundred call-seconds per hour (CCS) or call-minutes per hour (min). Since there are 3600 call-seconds in an hour, one Erlang is equal to 36 CCS or 60 call min.

Traffic flow through a switching centre is defined as the product of the number of calls during a period of time and their average duration. Traffic flow, therefore, can be expressed by the equation

traffic flow = (number of calls) × (mean call-holding time)

For example, if 100 calls of an average duration of three minutes are generated during a period of one hour, by the subscribers connected to the input of an office or local exchange, then the traffic flow for the group equals 300 call-minutes, or 5 call-hours.

9.4.2 Erlangs and unit calls

The international, dimensionless unit of telephone traffic is called the *Erlang*, named after a Danish telephone engineer, A.K. Erlang. One Erlang represents a circuit occupied for one hour; thus

$$1 \text{ Erlang} = 1 \text{ call-hour/hour} \qquad \qquad \ldots (9.4)$$

The number of Erlangs per busy hour may be calculated as follows (call-holding time expressed in hours):

$$\text{Erlangs} = (\text{calls/busy hour}) \times (\text{mean call-holding time})$$

For example, consider a connection established at 9.00 am between a central computer and a data terminal. Assuming that the connection was maintained continuously and data was transferred at say, a rate of 1200 bps, determine the amount of traffic, in Erlangs, transferred over the established connection between 9.00 am and 9.45 am:

$$\therefore \text{ Traffic} = (1 \text{ call}) (45 \text{ min}) (= 0.75 \text{hr}) = 0.75 \text{ Erlangs}$$

Note that the data rate is immaterial; the traffic is based solely on the call-holding time which, in this case, is 45 minutes or 0.75 hours.

When the call-holding time is expressed in seconds, the resulting traffic unit is the unit call or its synonymous terms hundred-call-seconds or centum-call-seconds (CCS), expressed as:

$$\text{CCS} = (\text{calls/busy hour})(\text{mean call-holding time})/100$$

Because there are 3600 seconds in an hour, the relationship between Erlangs and CCS is:

$$1 \text{ Erlang} = 36 \text{ CCS} \qquad \qquad \ldots (9.5)$$

CCS is the traffic unit employed mainly in North America, but, traffic expressed in Erlangs is generally more useful and provides direct information about the traffic, i.e.

- The Erlangs per channel represents its efficiency; that is, the proportion of the hour during which the channel is occupied.
- Traffic expressed in Erlangs designates the average number of calls in progress simultaneously during a period of one hour.
- Erlang figures represent the total time, expressed in hours, to carry all calls.

For example, if a group of 30 trunks are required to carry 12 Erlangs of traffic during the busy hour, what is the efficiency of this trunk group?

$$12 \text{ Erlangs}/30 \text{ trunks} = 0.40 \text{ Erlangs per trunk}$$
$$\therefore \qquad \text{Trunk-group efficiency} = 0.40 = 40\%$$

9.5 Grade of service

Grade of service (GOS) is a measure of the probability that a percentage of the offered traffic will be blocked or delayed. Grade of service, therefore, involves not only the ability of a system to interconnect subscribers, but the rapidity with which the interconnections are made. As such, grade of service is commonly expressed as the fraction of calls or demands that fail to receive immediate service (blocked calls), or the fraction of calls that are forced to wait longer than a given time for service (delayed calls).

The assessment of the grade of service provided to the user, and the determination of facilities required to provide a desirable grade of service, are based on mathematical formulas derived from statistics and the laws of probability. If a system is engineered on the basis of the fraction of calls blocked, then it is said to be engineered on a blocking basis. Blocking can occur if all facilities are occupied when a demand is originated or, in the case where several facilities must simultaneously be connected, a matching of idle facilities cannot be made even though only certain facilities are idle in each group. Some areas where engineering is done on the basis of blocking criteria are the dimensioning of switching matrices and inter-office trunk groups. Another basis for setting a service standard is the fraction of calls delayed before setting up, longer than an acceptable time. The delay encountered in providing voice channel after accessing a control channel for a subscriber is an example. As the number of common-control and time-sharing systems increases, so does the need for engineering on a delay basis.

9.5.1 Telephone traffic formulas

The assumption, blocked calls cleared, is the basis of a formula for obtaining the probability of loss, known as the *Erlang B* formula. The alternative is for some calls upon finding no channel available, to wait until one becomes idle, at which time the channel is seized and then held for the full holding time. This assumption, *blocked calls delayed*, is the basis of the *Erlang C* formula. Between these extremes is an intermediate assumption called *blocked calls held*, which are the basis of the so-called *binomial* and *Poisson* formulas. In this last case an offered call, upon finding no channel idle, waits for an interval of time exactly equal to its holding time and then disappears from the system. If a trunk becomes idle while the call is waiting, the trunk is seized and occupied for the portion of the holding time remaining. This is known as the lost-calls-held assumption to distinguish it from the other assumptions used in the Erlang formulas. We shall not discuss any of these formulas here, except to note that a mobile radiotelephone service is more closely described by the blocked-calls-cleared (Erlang B) formula when considering the voice channels.

Nevertheless, it needs to be noted that for the control channels (in an FDMA arrangement) a different situation applies. Firstly the messages (signalling) are very short, there is only a finite number of subscribers in a cell, and any blocked calls are cleared. An alternative traffic formula known as the *Engset* formula applies. Figure 9.2 show the result of calculating the likelihood of blocking (blocking probability) versus the offered traffic (Erlangs) for a three channel scheme. One notes how many more users than control channels can be accepted, and in most cases any control channel restriction is transparent to a subscriber.

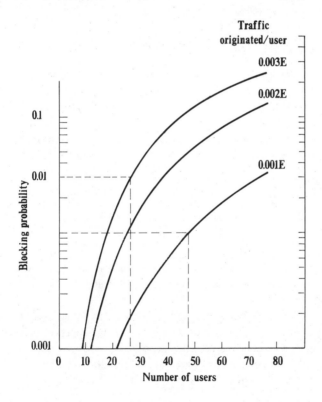

Figure 9.2   The probability of a cell being denied (blocked) in a three channel trunked system versus the number of users offering low calling rates (in Erlangs)

However, with regard to the voice or message channel the subscribers are likely to be making long calls, i.e. of the order of minutes. In a large cell the traffic throughout could amount to several Erlangs. One can now apply the Erlang B formula to find the offered traffic which meets a specified grade of service in a group of N channels.

9.5.2   Activity in a cell

Tables of results of the Erlang B formula, which list the grade of service applicable for an offered traffic with N voice channels, are given for example in the books by Boucher, referenced as further reading on this subject.

Firstly, we need to agree on some average calling rate for a subscriber and his holding time. Typical figures are six calls a day of 150 sec (2.5 min) duration. Hence traffic flow per subscriber is

$$\frac{6 \times 150}{60 \times 60} \times \frac{1}{24} = 0.0104 \text{ Erlangs}$$

Because our subscriber is unlikely to be operating over a 24-hour day, his traffic flow will tend to bunch up and the more likely traffic generated during the *busy hour* is, say, 0.03 Erlangs.

In any given cell we will have many (average) subscribers. How many is determined by the traffic capacity of the cell, or base station, assuming n_c channels and a certain grade of service, determined by the Erlang B formula; therefore

No. of subscribers per cell (BS)   =

$$\frac{\text{traffic capacity of base station}}{\text{calling rate of subscriber}} \qquad \ldots(9.6)$$

Table 9.1 lists the traffic capacity of a cell for a grade of service equal to 0.01, for the channel numbers n_c, which were listed in Table 3.1, for a 300 channel network.

Note that column (b) is the one shown as column (d) in Table 3.1. The interesting feature of the table, however, is that the number of subscribers per channel does not change very much with cluster size, when keeping GOS constant; in fact the figure of 30 used in (9.1) is quite a reasonable approximation. The benefit of a small cluster size is the minimization of handover in a given cluster area.

Table 9.1   Traffic capacity data for TACS type cellular plan
             (using data in Table 3.1) and GOS = 0.001

Cluster size N	No. of channels n_c	(a) Offered traffic GOS = 0.01	(b) No. of Subscribers per cell	(c) No. of Subscribers per CH
3	93	77.2	2583	28
4	69	55.2	1840	27
7	39	28.1	937	24
9	31	21.2	707	23
12	23	14.5	483	21

(a)   from Erlang B table
(b)   from (9.6)
(c)   column (b) divided by n_c .

Figure 9.3 illustrates further data for alternative grades of service and offered traffic (in Erlangs). The curves are worked out using Erlang B tables and (9.6), namely

$$\frac{\text{Subscribers}}{\text{per channel}} = \frac{\text{Traffic cap. of BS,} = n_c \text{ (assuming GOS = 0.01)}}{\text{Subscriber Erlang rate assumed} = 0.03} \times \frac{1}{n_c}$$

$$\ldots(9.7)$$

Cells with only a few channels can only handle a few subscribers; too many channels per BS has no clear benefits either. On the other hand, from a spectral efficiency viewpoint, it is clear that a measurement based on voice channels, definition (i), is a more precise definition.

9.6 Calculation of spectral efficiency

The definition: Channels per MHz per km^2 implies

$$\eta_s = \frac{\text{total no. of channels available}}{\text{total BW available} \times \text{cluster area}} \qquad \ldots(9.8)$$

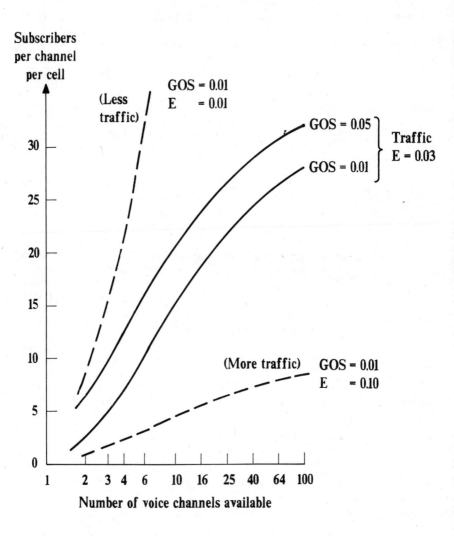

Figure 9.3　　The subscribers per channel per cell versus the total number of
　　　　　　　voice channels per cell for specific grades of service

Assuming a uniform cellular layout in clusters of N cells, each cell of which occupies an area S, and since

total BW available $= N \times n_c \times f_s$

where f_s is the bandwidth assigned to each subscriber

$$\therefore \eta_s = \frac{n_c}{BWS} = \frac{1}{f_s NS} \qquad \ldots (9.9)$$

This result indicates that narrowband (or low bit rate) technologies are best, small cluster sizes N are desirable, together with small cells. Note that, though η_s is termed an efficiency, by definition it is a number much greater than one, being the number of voice channels per MHz of allocated spectrum and the ground area being served.

9.6.1 Conventional FDMA

Here f_s refers to an individual voice channel, whether digital or analog modulation. In the digital case, we write

$$f_s = \frac{f_b}{mK} + 2\Delta f \qquad \ldots (9.10)$$

here f_b = voice coding bit rate
        m = transmission efficiency
        K = FEC rate as a fraction < 1
        Δf = hardware frequency error.

To minimize f_s,   need low f_b
                  m   high
                  K   = 1
                  Δf minimum

f_b and m were discussed in Chapter 6. For GMSK, measurement of the 60 dB adjacent channel performance shows that with BT = 0.3, m ~ 1.35; which implies a 16 kHz channel needed for 22 kbps speech, which allows for the code correction bits.

Alternatively, $\pi/4$ DPSK offers m ~ 1.62, which allows 48 kbps in 30 kHz, which is why it is being considered for the ADC system.

Turning to cluster size N, the co-channel interference dependence was introduced in Chapter 3. Recalling (3.1)

$$N = \frac{1}{3}\left(\frac{D}{R}\right)^2$$

where R = cell radius and D = reuse distance, we found that the carrier-to-co-channel interference ratio C_i could be written as

$$C_i = 1.5N^2 \qquad\qquad\qquad \ldots (9.11)$$

assuming an inverse fourth power propagation law. More precise calculations of C_i versus N are possible, but do not affect the results below significantly.

To a reasonable approximation therefore we can put, by combining the results,

$$\eta_s = \frac{m.K}{f_b} \times \left(\frac{3}{2C_i}\right)^{\frac{1}{2}} \times \frac{1}{S} \qquad\qquad \ldots (9.12)$$

where $\Delta f \to 0$.

As a further helpful approximation, let the factor mK balance out the number root 3/2, and since $S = \pi R^2$, where R = cell radius,

$$\therefore \quad \eta_s = \frac{1}{f_b} \times \frac{1}{(C_i)^{\frac{1}{2}}} \times \frac{1}{\pi R^2} \qquad\qquad \ldots (9.13)$$

The result is interesting in the sense that it is the carrier-to-co-channel interference ratio which should be minimized, and to some extent the number of cells in a cluster is immaterial.

The other factor which must be kept in mind, however, is the relation between the FEC and C_i ratio, since more correction may be needed to sustain a lower C_i ratio. The whole matter is complicated by the fading statistics of the wanted and interfering signals.

9.7 Multi-access efficiency factor

Provided that the channel discussed above can be fully loaded with voice traffic, then the efficiency, in Erlangs per MHz per km^2, is almost equal to the factor η_s above. To load up each channel a multi-access scheme must be employed, and one assumes that there are many more users than channels. How efficient is this access facility? One approach is to multiply η_s by a term η_a, where

$$\eta_a = \frac{\text{Total time/bandwidth product devoted to the voice channels}}{\text{Total time/bandwidth product devoted to the system}} \qquad \ldots (9.14)$$

$$\therefore \quad \text{overall efficiency} \quad \eta_o = \eta_s \times \eta_a$$

For *FDMA*, this has been discussed in section 6.2 in relation to the need for guard bands between channels, and also the need for control channels in the arrangement, see Figure 4.2.

Clearly $\quad \eta_a \leq 1$

For *TDMA*, the efficiency will depend on how many bits in a time frame are dedicated to the message and how many are overhead. In narrowband TDMA systems, an FDMA breakup of the available spectrum also exists, see Figure 8.2.

For a TDMA channel group, one has

$$\eta_a = \frac{\text{(time slot duration for voice)} \times N_F}{\text{(frame duration)}} \qquad \ldots (9.15)$$

where
N_F = no. of time slots for voice transmission per frame.

To work out the access efficiency using bit numbers, i.e. using (9.15), requires taking out the parts of the message used for forward error correction, etc.; but in the first instance, let

$\tau \quad = \quad$ time slot duration of the message
$T_F \quad = \quad$ frame duration

$$\therefore \quad \eta_a \quad = \quad \frac{\tau . N_F}{T_F} \qquad \ldots (9.16)$$

If one takes GSM (Chapter 8) as an example, then we have

Time slot period	=	576.9 µs
Frame period	=	4.615 ms
No. of users per frame	=	8

Hence $\quad \eta_a \quad = \quad \dfrac{0.5769 \times 8}{4.615} \quad = \quad$ 1.0 or 100%

Unfortunately, much of each time slot is taken up with error protection bits, tail bits, training sequence and a guard space, so that a more accurate indication of the access efficiency comes from the alternate formula

$$\eta_a = \frac{\text{(voice channel bps)} \times N_F}{\text{(Group bandwidth bps)}} \qquad \ldots (9.17)$$

Again taking GSM as an example, and using (9.17) we have

voice channel bit rate	=	13 kbps
group bandwidth bit rate	=	270.833 kbps
No of channels per frame	=	8

$$\therefore \quad \eta_a \quad = \quad \frac{13 \times 8}{271} \quad = \quad 0.38 \text{ or } 38\%$$

For narrowband TDMA, as explained, more spectrum is available in the other channel groups on an FDMA principle. Therefore, the access efficiency must also be modified by the term

$$\eta'_a \quad = \quad \frac{\text{(bandwidth for a frame)} \times \text{no. of bands}}{\text{(total BW of system)}} \qquad \dots (9.18)$$

Let bandwidth per frame $\quad = \quad B_f$
no. of radio bands $\qquad = \quad N_c$

$$\therefore \quad \eta'_a \quad = \quad \frac{B_f \times N_c}{BW} \qquad \dots (9.19)$$

In GSM, $B_f = 200$ kHz, $N_c = 125$ and BW = 25 MHz (twice for duplex)

$$\therefore \quad \eta'_a \quad = \quad \frac{200 \times 125}{25,000} \quad = \quad 1$$

so no efficiency is lost through the narrowband TDMA format plan.

In TACS, $B_f = 25$ kHz, $N_c = 1000$, again BW = 25 MHz, so the access efficiency would appear to be unity. As we know several channels are allocated to control channels between the different operators, so the access efficiency is actually less than unity.

9.7.1 Overall efficiency

The overall efficiency is the product of spectral efficiency and access efficiency i.e.

$$\eta_o \quad = \quad \eta_s \times \eta_a \qquad \dots (9.20)$$

Because, as we found, η_a is often effectively one hundred percent, we can accept the overall efficiency, actually the number of voice channels as discussed after equation (9.9), is determined by

$$\eta_s$$

which is decided by the terms in equation (9.13).

The formula for η_s however, it will be recalled, is an approximation; it hides the loss of performance due to FEC. Also, in the case of narrowband TDMA the access efficiency is not unity, again due to all the overheads in the speech coding algorithm, the time slot and the frame itself, and appears to represent quite a serious loss of efficiency. However, the speech coding provides transmission security by optional levels of encryption and therefore represents additional service over and above plain speech. Nevertheless, the basic formula for η_s indicates the real limits of cellular radio. These are

(i) the voice communication bandwidth or bit rate. Here digital technology is now drawing level with analog.

(ii) the tolerable carrier to co-channel interference. A 6 dB change here can represent a 40% improvement in efficiency. This is the aim of digital technology.

(iii) Any minimizing of the cell size will show real benefit. Power control of the base station and mobile transmitter by digital control makes for smaller cell geometry.

Using TDMA avoids the problem of equipment centre frequency drift as compared to narrowband FDMA technology, but as the bit rate per channel is increased more irreducible BER is encountered on the radio path due to multipath. This irreducible BER is due to intersymbol interference (ISI) caused by multipath delay spread s. Experience has shown that ISI becomes noticeable when

$$\text{Signal bit rate} < 10 \text{ s} \qquad \ldots (9.21)$$

Thus in the case of the proposed ADC system we have $T_b = 1/48600 = 20.5$ μs, which is about ten times the average delay spread. For GSM, on the other hand, the signalling rate is 271 kbps, corresponding to a symbol period of 3.7 ms, which comes much closer to the delay spread value of say 2 μs. This is why 26 bits need to be dedicated to an in the time slot training period within the GSM frame format. Therefore, some of the efficiency of a wider band TDMA format has to be given over to extra overhead within the signalling.

Further reading

Boucher, J.R. (1990). *Cellular Radio Handbook,* Quantum Publishing Inc, USA

Boucher, J.R. (1988). *Voice Teletraffic Systems Engineering,* Artech House Inc, USA

Chuang, J.C.I. (1989). 'The effects of delay spread on 2-PSK, 4. PSK, 8-PSK, and 16- QAM in a portable radio environment', *IEEE Trans Veh Tech,* Vol 38, May, pp 43-45

Farr, R.E. (1988). *Telecommunications Traffic, Tariff and Costs*, Peter Peregrinus, IEE Press, UK

Hatfield, D.N. (1977). 'Measures of spectral efficiency in land mobile radio', *IEEE Trans EMC*, Vol 19, Aug, pp 266-268

Hummuda, H., McGeeham, J.P. and Bateman, A. (1988). 'Spectral efficiency of cellular land mobile radio systems', *IEEE Veh Tech Conference no 38*, pp 616-622

Lee, W.C.Y. (1989). 'Spectrum efficiency in cellular', *IEEE Trans Veh Tech*, Vol 38, May, pp 69-75

Murota, K. (1985). 'Spectrum efficiency of GMSK land mobile radio', *IEEE Trans Veh Tech*, Vol 34, May, pp 69-75

Raith, K. and Uddenfeldt, J. (1991). 'Capacity of digital cellular TDMA systems', *IEEE Trans Veh Tech*, Vol 40, May, pp 323-331

Appendix

Summary of current world systems

The commonality within cellular radio systems was highlighted in the introductory chapter. The various systems currently operating and being planned to operate, globally, have been referred to at appropriate places in the text. Here we list a summary of the main parameters of the systems, shown as Table A.1.

Each system is known by an abbreviation, listed here. The number of subscribers refers to data taken from various 1991 global sources. The cellular principle parameters are those defined in the text. Table A.1 is by no means exhaustive; however, it does illustrate how feature differences make for wide incompatibility.

NMT   -   Nordic mobile telephone (followed by a number referring to the frequency band).

AMPS   -   Advanced mobile phone system (USA).

TACS   -   Total access communication systems (UK).

C-NET   –   An analog network, specific to Germany.

GSM   -   Global system for mobile communications - specified by CEPT (Committee of European Posts and Telecommunications).

E-TACS -   Extended TACS, offering more channels by additional frequency assignment.

JTACS   -   Japanese total access communications system, similar to TACS.

JDC   -   Japanese digital cellular; a narrowband TDMA system.

ADC   -   American digital cellular (similar to JDC); works within the AMPS frequency plan.

System Name	Start Date	Frequency Band MHz BS$_{Tx}$	Frequency Band MHz MS$_{Tx}$	Mode and Modulation	Channel Bandwidth kHz	Data Rate kbps	Cluster Size	No. of Channels	Handover System	Approx. no. of Countries	Approx. no. subscribers (millions)
NMT450	1981	463-467.5	453-457.5	FDMA FM ±4.7 kHz	25 20	1.2	60° sector	180 225	C/N at BS	14	1.5
AMPS	1983	869-894	824-849	FM±12 kHz	30	10.0	7 12	832	C/N at BS	37	9
TACS (ETACS has extended BW)	1985	935-950 917-933	890-905 872-888	FM ±9.5 kHz	25	8.0	7 12	1000 1640	C/N at BS	21	2.5
NMT900	1986	935-960	890-915	FM ±4.7 kHz	25 12.5	1.2	7	1000 2000	C/N at BS	8	0.5
J-TACS J-NAMPS	1979 1988	925-940	870-885	FM±5 kHz ±2.5 kHz	25 12.5	0.3 2.4	14	600 1200 (2400)	C/I value	1	2
C-Net	1985	461-466	451-456	FDMA ±4 kHz	20	5.28	–	222	Propagation delay at BS	3	0.5
GSM (occupied by TACS)	1992	950-960 (935-950)	905-915 (890-905)	TDMA (GMSK)	200	271	4	(75 radio channels x 8 TDMA) = 600	BS interrogation	24 (based on M.o.U)	(starting)

Table A.1 Parameters describing current available cellular radio systems

Index

Access burst 24, 185
  channel 24
  grant channel 183
Adaptive differential PCM 149
Adaptor predictive coding 149
ADC 60, 139, 157–9, 186–9, 203, 207, 209
ADC features 171
ADC speech coder 157
Additional loss factor 50
Additive white Gaussian noise 28, 142
Adjacent channel 7, 130
  channel interference 70, 74
Air interface 174
Aircraft cellular system 66, 188
Allocation 7, 60, 210
AMPS 28, 60, 74, 98, 123, 184, 209
Antenna aperture 40
  combination unit 63
  gain 8, 41
Area identification 84
Automatic repeat request 26, 103
Assigned bandwidth 120
Associated control channel 183
Authentication 14, 23, 166
Authentication centre 14, 23

Balanced mixer 125
Base (station) colour code 23, 73
Base sites 62
Base station 3, 8, 15, 79, 180
  application part 172
  controller 13, 172
  interface 172
  system 62, 79, 173, 191
BCH coding 24, 28
Billing centre 14, 166, 193
Binary modulation 126, 139
Bit error probability 141
Bit error rate 21, 53, 140–145, 207
Block codes 28

Boltzmann's constant 43
Broadcast control channel 176, 183
BT product 136, 203
Building penetration loss 53
Burst transmission 163, 177, 185

Call holding time 195
Call set-up 4, 92
Capture effect of FM 20
Carrier phase 121
Carrier-to-interference ratio 69, 204
Carrier-to-noise ratio 20, 41, 43, 140
Carson bandwidth 124
CCIR 46, 48, 51
CCITT 26, 30, 32, 80, 147, 149, 173
Cell allocation 7, 67
  broadcast channel 182
  clusters 8, 13, 67
  splitting 72
Cellnet 62
Cellular features 35
Cellular network 14, 78, 86
CEPT 165, 209
Channel bandwidth 122, 195, 210
  group 161
  number 60, 192
  spacing 60
Cipher key 26
City area 194
Cluster size 12, 67–73, 195
Clutter factor 51
Co-channel interference 69, 77–9, 195, 204
Code division multiple access 159
Codebook vocoders 156
Coherence bandwidth 26, 116
Coherent frequency exchange keying 130
Coil loading 40
Colour code 23
Common control channel 181
Control channel 22, 61, 65, 75, 82, 178
  messages 89

Convolutional code 28, 154, 157
Country code 32, 85, 172
Coverage prediction 53–57
Cyclic redundancy check 28, 103

Data communication networks 100
Data field 88
  link layer 168
  packet 100
  rate 210
  specific network 100
Dedicated control channel 24, 83, 183
Delay spread 26, 116, 207
Delta modulation 150
Detector 144
Differential PSK 29, 138, 188, 203
Digital colour code 23, 73, 88, 91, 188
Diversity reception 114
Doppler shift 105, 111, 177
Double sideband 125
Downlink 8, 175, 180, 182
Dual-tone multi-frequency 2-3
Duopoly 60
Duplex operation 3, 59, 164
Duplexers 164
Dynamic multipath 107

Effective intrinsic radiated power 42
Electromagnetic wave 37
Electronic serial number 34, 91
Energy per bit 29, 141
Equalizer 179
Equipment identity register 14
Erlang 196–200
Error correction 25–30, 154, 175
  correction rate 203
  protection 154, 157, 188
ETSI 165, 173
Extended TACS 61, 209

Fading envelope 107, 112
Fast associated control channel 184
Fast fading 106, 112
Fast FSK 65, 74, 129, 160
FDMA 17, 59–62, 119, 159, 162, 203
Filtered FSK 134
FM demodulation 20, 123

deviation 83, 121
  index 121–123
  pre-emphasis 123, 155
  threshold 20
  transmit 123
Forward control channel 22, 83, 87
  error correction 22, 27–29, 145
  path 4, 60
  voice channel 4, 22, 87
Four-level FSK 131
Fourth power propagation law 48, 66
Frame arrangement 27, 162
  number 176
  period 162
Free space loss factor 42
Frequency agility 12, 182
  allocation 17, 59
  deviation 83
  hopping 182
  modulation 18–20, 119, 121
  selective fading 105, 115, 117
  shift keying 21, 128
Fresnel zones 55
Full-rate channel 153
  speech 153
  traffic channel 184

Gaussian minimum shift keying 135, 170
Grade of service 64, 198, 200
Grazing angle 46
GSM 159, 165–185, 209
GSM channel coding 176–179
  channels 183
  features 166, 171
  fixed network 172
  frame 176, 181, 205
  interfaces 172
  logical channels 183
  mobile 169
  modulation 137, 178, 203
  OSI model 167
  parameters 175
  power levels 174, 181
  recommendations 167, 170
  spectrum allocation 83, 210
  speech coder 153–155

speech decoder 155
Guard (space) 163, 178

Half-rate traffic channel 184
Half-wave dipole 38–40
Handover 11, 76, 96, 182
Handportable unit 76
Hexagonal shape 66–76
High level data link control 23, 27
Home location register 14, 78, 85, 173
  mobile switching centre 166
Hyperframe 176

Identification 166
In-flight cellular 66
Interleaver 178
Intersymbol interference 135, 207
Irreducible BER 22, 142
ISDN 31, 165
ITU 17

Japanese digital cellular 159, 189, 209
  TACS system 209

Knife edge model 54

Linear predictive coding 152–157
Local exchange 2
Location area 75
  area code 88
Long-term predictor 153

Macrocells 11
Manchester encoding 84
Mean opinion score 147
Message bandwidth 122, 195
Microcells 11, 73
MIN code 14, 33, 84, 91, 166, 173
Minimum shift keying 119, 129, 134
Mobile application part 173
  attenuation code 89
  channel number 60, 89
  class 10, 180
  country code 33, 85
  location 75
  network code 85
  numbering 31, 85

power control 77, 79
station 1–6
station number 5, 32, 171
subscriber numbers 6, 31, 173
switching centre 13, 78, 80, 173, 191
Modulation 18, 119, 210
Modulation bandwidth 121
  efficiency 178
  index 121, 128
  sidebands 119
Multi-access efficiency 202
Multi-level modulation 140
Multiframe 176
Multipath envelope functions 109, 113
  fading 70, 104
  level crossing rate 113
  power spectral density 112
  propagation 104
  reception 26, 142
  simulation 111
  testing 118
Multiple access 62, 162, 204
  cell sites 66–76
  cells 66
Multi-pulse vocoder 152

Narrowband AMPS 123
  multipath 105
  TDMA 161
National (significant) number 31
  mobile station identification 31
Neighbouring (adjacent) cell 65, 68
Network destination code 31
  element 78
  management 171, 193
  termination unit 101
NMT 8, 23, 74, 98, 123, 209
No. of channels per cell 120, 194
Noise figure 43, 146
  power 141
  temperature 43
Normal burst 185
Number assignment module 34
Number of users 10, 71, 194
Numbering plans 30–34
Off-set PSK 134
Offered traffic 196

Open system interconnection 167
Operations and maintenance centre 16, 191
OSI layers 168
Other services 80
Overall efficiency 206
Overhead message 87
Overlaid cell 74

Packet signalling 102
Paging in cellular 24, 86, 93, 183
Paknet 100
Path loss 48
Parity field 88
Personal identification number 166
Phasor 18
Phase quadrature 18
    shift keying 125
Physical layer 168, 176
Polarization 37
Population 194
Power level 10, 89
    ramp 181
Private mobile radio 11, 53, 82, 106
Propagation loss 42, 48
PSTN 14, 26, 30, 172
Public telecommunications operator 196
Pulse code modulation 104, 147

Quadrature phase 18
    shift keying 129–135, 142

Radiation power density 38, 42
Radio antenna 3, 38–42
    channels 4, 196
    conferences 16
    link protocol 168, 180
    range 9
    system entity 174
    waves 18, 37
Radiotelephone 1–6
Raleigh distribution 109
Random access channel 183
Rayleigh fading 110
Receiver 4, 8, 79
Reflection of radio waves 45
Registration 76, 90

RELP 152, 160
Repeat cell 8, 67
    distance 8, 68
Repeater 11, 53
Reserved for future use (bits) 88
Residual excitation 152
Return path 4, 60
Reuse 8, 68, 71
    distance 8, 68
    ratio 68
Reverse control channel 22, 90, 92
    voice channel 4, 22
Roaming 5, 34, 8

SAT colour code 84, 89, 95
Scattering model 108
    profile 117
Sectorization 67, 73
Security 166, 207
Send 5, 92
Serial number 34
Service area 8, 73, 90
    provider 14
Services 80, 166
Shannon limit 29
Shannon's formula 145
Short message service support 164
Signal bandwidth 119
Signal-to-noise ratio 20, 41, 44, 147
Signalling activities 24, 65
    tone 84, 95
SIM card 166
Site engineering 62
Slow control channel 183
    fading 106
Spectral efficiency 191, 195, 201–204
Spectrum allocation 61
    dividing filters 62
Stand-alone control channel 184
Standing wave pattern 106
Station class mark 91
Stockholm plan 74
Stored program control 31
Subscriber interface module 166, 172
    number 32
Subscribers 6, 10, 191
    per channel 202

Supervisory audio tone 23, 73, 84, 89, 95
Switching offices, see MSC, 192
Symbol 27, 126
Synchronization burst 178, 180, 185
Synthesizer 4, 11, 169

TACS 60, 72, 98, 123, 194, 206, 209
Tail bits 89, 177
Tamed FM 135
TDMA 17, 28, 62, 72, 83, 119, 159–164, 205
Telephone number 2–5
    traffic formula 198, 201
Terrestrial propagation 47
Time delay 105, 117
    slot 28, 162, 173, 176, 182–189
    slot number 175, 182
Timing advance 180
Toll quality speech 147
Traffic capacity 196, 201
    channel 176, 185
    frame 24, 162
Training sequence 179, 185
Transceiver 169
Transit switching centres 14, 86
Transmit power 10, 79, 180
Transmitter 4, 8
    emission mask 160

Trellis coding 145
    diagram 137
Trunk code 5, 41
Trunking 63, 82
    gain 64, 194
Type approval 179

Uplink 8, 175, 180, 182
User bandwidth 120

Vector sum vocoder 157
Vehicle velocity 21, 111
Vertical services 82, 168
Visitor location register 14, 78, 85, 173
Viterbi coding 27
Vocoders 148
Vodafone 57, 61, 75, 86
Voice analysis 15
    attenuation code 89
    channels 9, 23, 64, 82
    model 150
    synthesis 151
VSWR 62
Waveform coders 148
Wavelength 37
Wideband multipath 106
    TDMA 161

Yagi antenna 40